Multiplexing

Edited by Somayeh Mohammady

Published in London, United Kingdom

IntechOpen

Supporting open minds since 2005

Multiplexing
http://dx.doi.org/10.5772/intechopen.78513
Edited by Somayeh Mohammady

Contributors
Kasturi Vasudevan, Shivani Singh, A. Phani Kumar Reddy, Nandalal Vijayakumar, Sumalatha M S, Huacheng Zeng, Pedram Kheirkhah Sangdeh, Younes Zouine, Zhour Madini, Meena Dasan, Fredy Francis, Sarath K T, Dipin E, Srinivas T, Somayeh Mohammady

First published in London, United Kingdom, 2019 by IntechOpen
IntechOpen is the global imprint of INTECHOPEN LIMITED, registered in England and Wales, registration number: 11086078, The Shard, 25th floor, 32 London Bridge Street
London, SE19SG - United Kingdom
Printed in Croatia

British Library Cataloguing-in-Publication Data
A catalogue record for this book is available from the British Library

Additional hard and PDF copies can be obtained from orders@intechopen.com

Multiplexing
Edited by Somayeh Mohammady
p. cm.
Print ISBN 978-1-78984-568-6
Online ISBN 978-1-78984-569-3
eBook (PDF) ISBN 978-1-83881-868-5

For EU product safety concerns:
IN TECH d.o.o., Prolaz Marije Krucifikse Kozulić 3, 51000 Rijeka, Croatia, info@intechopen.com or visit our website at intechopen.com.

Meet the editor

Somayeh was born in 1982 in Shemiran, Tehran. She went to secondary school of Dr. Mahmoud Hessabi in Shemiran, Tehran. She finished her high school in 1998 specializing in Math and Physics in Kherad private school managed by Dr. Esmat Mohammadi and received her diplomas in 1999 in Math, Physics, and Computer Science from Tolou pre-college institute, and TVTO, Karaj. She tool trained with Hormoz Haghighi at Saba Animation broadcasting institute. Somayeh Mohammady, BSc, MSc, PhD, is from Shemiran, Tehran. She earned a BSc in Electronic Engineering from Azad University of Naein, where she built and assembled two line-following robots using infrared sensors and DC/AC motors. She earned her graduate degrees in Electronic Engineering from University Putra Malaysia (UPM). For her MSc, she worked under the supervision of Dr. M. Nizar Hamidon on Power Amplifier (PA) Linearization, and for her PhD, she worked under the supervision of Prof. Roslina M. Sidenk and Dr. Nasri Sulaiman on Peak-to-Average Power Ratio (PAPR) reduction/Crest Factor Reduction (CFR). Dr. Mohammady has experience working as an engineer for Tabesh Tablo, Teta, and the Symmid Corporation. She also works as a postdoc researcher and teaching assistant. She is with the CONNECT research group in the Department of Electronic Engineering at Maynooth University (MU), Maynooth, Kildare, Ireland. Dr. Mohammady has authored 22 publications, and co-authored more than 20 publications worldwide, including conference and journal articles, book chapters, and books.

Contents

Preface

This book examines the concept of multiplexing through the lens of telecommunication electronics, digital signal processing, and robotics. Topics highlighted include multiple-input and multiple-output (MIMO) and 5G-related applications. It is fascinating to learn that the basic idea of multiplexing, which dates back to the time of telegraph systems, is still very valuable today and is, in fact, the basis for multicarrier systems such as Long-Term Evolution (LTE) communication, 5G technologies, optic systems, and the Internet of Things (IoT), among many others. However, it should be noted that multiplexing is not limited to only these areas, but is widely used in many fields. Multiplexing is mainly used to share resources and enable simultaneous processes, and these processes and resources may vary from one field to another.

To start, this work presents a brief history of multiplexing, including relevant definitions and an overview of the components used in multiplexing systems. Then, it goes on to discuss some of the most practical applications of multiplexing, such as Multiple-Input, Multiple-Output Orthogonal Frequency-Division Multiplexing (MIMO-OFDM), Direct-Sequence Optical Code-Division Multiple-Access (DS-OCDMA) systems, and optical multiplexing.

I want to thank my parents, Mrs. Shahin Mohammady and Mr. MohammadAli Mohammady, my husband, Dr. Pooria Varahram, and my children, Daniel and Denise, for all their compassion, consideration, and help.

I also want to thank my supervisors, Prof. Ronan Farrell and Dr. John Dooley, and my colleagues Joanna O'Grady, James Kinsella, and all members of the Department of Electronic Engineering at Maynooth University, for all their care, support, kindness, forgiveness, guidance, and advice.

This publication is a result of research conducted with the financial support of Science Foundation Ireland (SFI) and is co-funded under the European Regional Development Fund under Grant Number 13/RC/2077.

Somayeh Mohammady
National University of Ireland Maynooth,
Ireland

Section 1

Introduction

Chapter 1

Introductory Chapter: Multiplexing History - How It Applies to Current Technologies

Somayeh Mohammady and Pooria Varahram

1. Definition of multiplexing

It is fascinating to know that the multiplexing can be employed and defined from different points of views and different disciplines. For example, Dr. E. Julius Dasch, former manager of the NASA National Space Grant Program, defines multiplexing on his dictionary book [1] as:

> *...the simultaneous transmission of different data from a spacecraft using a single channel. The data stream is separated into frames that carry codes for different information, such as temperatures, pressures, and the state of on-board computers.*

In a media and communication dictionary book [2], multiplexing is mentioned as:

> *Digital television transmission allows for multiplexing, whereby multiple channels are bundled together and sent simultaneously in a single stream of data.*

In a computer science dictionary book [3], multiplexing is described as:

> *...the process of combining multiple messages simultaneously on the same physical or logical transmission medium. There are two main types: time division multiplexing (TDM) and frequency division multiplexing (FDM).*

The list continues for the Internet, optic engineering, graphics, phytography, laser and photonics geology, earth sciences, and many more.

2. Short history of multiplexing

The history of multiplexing goes back to the 1800s, when Samuel Morse developed his telegraph system which enabled long-distance communications [4]. Later in 1874, Thomas Edison invented diplexing to transmit two individual messages over one line at the same time [5]. Later in 1894 and the 1930s, time-division multiplexing (TDM) and frequency division multiplexing (FDM) came into existence [6, 7].

3. Application of multiplexing in current technology

One of the expectations about 5G technologies is to support enormous capacity, approximately 1000 times devices per squared kilometer [8]. In order to satisfy this requirement, several technologies have been suggested and developed, and one of the most attractive approaches is known as massive multiple-input multiple-output (MIMO), and that is where, for example, spatial multiplexing comes to use [9]. From signal processing side, for instance, employing multiple carriers all the way to arranging antennas and network management, multiplexing technique provides accessing resources by dividing and sharing it among users. In the other side, de-multiplexing has to be applied at the receiver side to inverse all the processes and extract the information sent.

One particular example for application of multiplexing is seen in orthogonal frequency division multiplexing (OFDM) systems [10]. The signal is spread between different subcarriers, and the frequency bandwidth is efficiently used [11].

4. Conclusion

This chapter is an introduction to the book titled *Multiplexing*. A variety of definitions of multiplexing from different points of view are presented, and a short history of the origin of multiplexing is briefly discussed. The potential and application for existing and future technologies are also discussed.

Acknowledgements

This publication has emanated from research conducted with the financial support of Science Foundation Ireland (SFI) and is co-funded under the European Regional Development Fund under Grant Number 13/RC/2077.

Author details

Somayeh Mohammady[1*] and Pooria Varahram[2]

1 National University of Ireland Maynooth, County Kildare, Ireland

2 Benetel Ltd., Dublin, Ireland

*Address all correspondence to: somayeh.mohammady@mu.ie

References

[1] O'Meara S, Dasch EJ. A Dictionary of Space Exploration. England, United Kingdom: Oxford University Press; 2018. DOI: 10.1093/acref/9780191842764.001.0001. eISBN: 9780191842764

[2] Chandler D, Munday R. A Dictionary of Media and Communication. England, United Kingdom: Oxford University Press; 2016. DOI: 10.1093/acref/9780191800986.001.0001. eISBN: 9780191800986

[3] Butterfield A, Ngondi GE. A Dictionary of Computer Science. England, United Kingdom: Oxford University Press; 2016. DOI: 10.1093/acref/9780199688975.001.0001. Print Publication. ISBN-13: 9780199688975. eISBN: 9780191768125

[4] Coe L. The Telegraph: A History of Morse's Invention and Its Predecessors in the United States. England, United Kingdom: McFarland Inc.; 2003. ISBN: 0786418087, 9780786418084

[5] Anderson KC, Edison T. The Importance of. United States: Lucent Books; 1994. ISBN: 1560060417, 9781560060413

[6] Russell J, Cohn R, Time-Division Multiplexing, Book on Demand (BOD). 2012. Published online. ISBN: 5512182293, 9785512182291

[7] Beyda WJ. Data Communications: From Basics to Broadband. United States: Prentice Hall; 1996. ISBN: 0133669238, 9780133669237

[8] Luo FL, Zhang CJ. Signal Processing for 5G: Algorithms and Implementations. United States: IEEE/John Wiley & Sons; 2016. ISBN: 1119116465, 9781119116462

[9] Darbai F, Stewart RW, Glover IA. MIMO channel modelling. In: Signal Processing. Rijeka, Croatia: IntechOpen; 2010. DOI: 10.5772/8530

[10] Mohammady S, Sulaiman N, Sidek RM, Varahram P, Nizar Hamidon M. FPGA implementation of inverse fast Fourier transform in orthogonal frequency division multiplexing systems. In: Fourier Transform. Rijeka, Croatia: IntechOpen; 2011. DOI: 10.5772/36664

[11] Varahram P, Mohammady S, Hamidon MN, Sidek RM, Khatun S. Demonstration of a power amplifier linearization based on digital predistortion in mobile Wimax application. In: Advanced Microwave and Millimeter Wave Technologies. Rijeka, Croatia: IntechOpen; 2010. DOI: 10.5772/8753

Section 2

Multiplexing Basics

Chapter 2

Multiplexing

Vijayakumar Nandalal and M.S. Sumalatha

Abstract

In any communication system that is either digital or analog, we need a communication channel for transmission. This channel can be a wired or a wireless link. It is not practical to allocate individual channels for each user. Therefore a group of signals are combined together and sent over a common channel. For this we use multiplexers. A multiplexer is a device that allows digital information from several sources to be routed onto a single line for transmission to a single destination. A demultiplexer does the reverse operation of multiplexing. It takes digital information from a single line and distributes it to a given number of output lines.

Keywords: frequency division multiplexing, time division multiplexing, code division multiplexing, wave length division multiplexing, orthogonal frequency division multiplexing, amplitude shift keying, frequency shift keying, phase shift keying

1. Introduction

Multiplexing is the process of transmission of information from more than one source into a single signal over a shared medium. We can be able to multiplex analog or digital signal. If analog signals are multiplexed, then this type of multiplexer is called analog multiplexer. If digital signals are multiplexed, then this type of multiplexer is called digital multiplexer. The advantage of multiplexing is that we can transmit a large number of signals to a single medium. This channel can be a physical medium like a coaxial, metallic conductor or a wireless link and will have to handle multiple signals at a time. Thus the cost of transmission can be reduced.

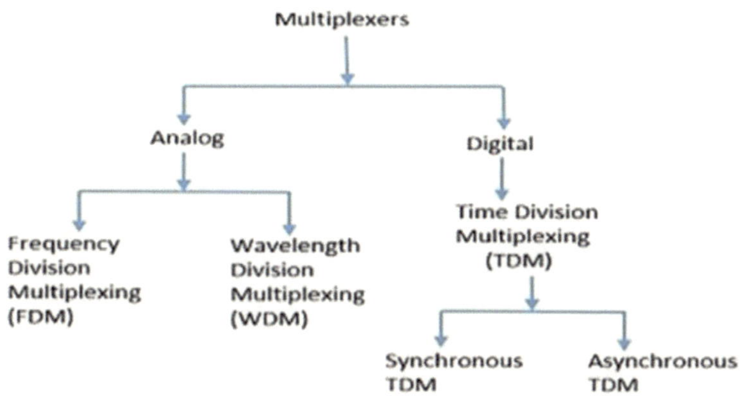

Figure 1.
Classification of multiplexing techniques.

Even though the transmission occurs on the same channel, they do not necessarily occur at the same instant. In general multiplexing is a technique in which several message signals are combined into a composite signal so that these can be transmitted over a common channel. In order to transmit various signals over the same channel, it is essential to keep the signals apart to avoid the interference between them, and then it can be easily separated at the receiving end.

Domains in which multiplexing can be accomplished are time, phase, frequency wavelength, etc. Multiplexing circuits are called multiplexer or MUX.

1.1 Types of multiplexing

Multiplexers are mainly classified as shown in **Figure 1**. Analog multiplexing and digital multiplexing are the major classification.

2. Analog multiplexing

The most commonly used analog multiplexing techniques are frequency division multiplexing (FDM) and wavelength division multiplexing.

2.1 Frequency division multiplexing

Frequency division multiplexing [1–4] is a networking technique which combines many signals into a single one and then transmitted the combined signal through a common communication channel. In the receiver side, the opposite process is carried out which is known as demultiplexing which extracts the individual channel signals. Here the transmitter side performs multiplexing, and the receiver side performs demultiplexing. In FDM the total bandwidth available in a communication medium is divided into a series of nonoverlapping frequency bands. Each of these bands is used to carry a separate signal. In FDM all users use the same common channel at full time. But each of the users is allocated with different frequencies for transmission for avoiding the signal interference. Sometimes there is a possibility of cross talk because all the users use the transmission medium at the same time.

FDM is used for analog signal transmission. It does not need synchronization between the transmitter and receiver. Here a large number of signals can be transmitted simultaneously. It suffers the problem of cross talk, and intermodulation distortion may take place.

FDM is used in amplitude modulation (AM) and FM broadcasting, public telephone networks, and cable TV network systems. The allocation of frequency bands to different users is shown in **Figure 2**.

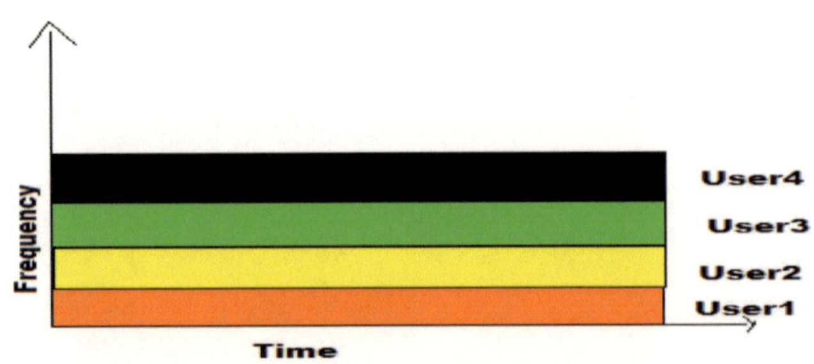

Figure 2.
Allocation of different frequencies to different users in FDM.

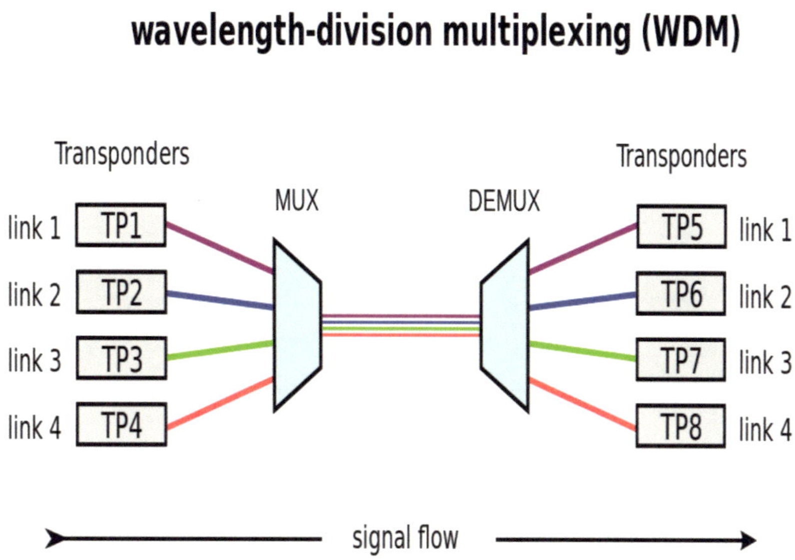

Figure 3.
Representation of WDM multiplexing.

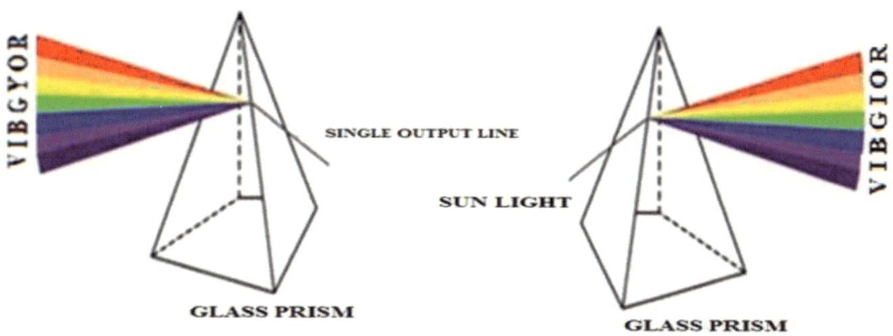

Figure 4.
Refraction of light through prism.

2.2 Wavelength division multiplexing

Fiber-optic communications require a different kind of multiplexer called a wavelength division multiplexer (WAD) [2, 4]. It is an analog multiplexing technique. It is designed for high data rate capability fiber cable. In this technique the bandwidth of the communication channel should be greater than the combined bandwidth of the individual channels. Here signals are converted to light signals; each light which has different wavelengths is transmitted through the same fiber cable. WDM transmission system divides the optical fiber bandwidth into a number of nonoverlapping optical wavelengths; these are referred to as WDM channels. WDM mixes all incoming signals having different wavelengths and are transmitted over a common channel. A demultiplexer does the reverse operation and separates the wavelengths. This multiplexing mechanism provides a much higher available transmission capacity.

Figure 3 shows the representation of WDM system that consists of both multiplexer and demultiplexer.

WDM multiplexing and demultiplexing are similar to the refraction of light through a prism as shown in **Figure 4**.

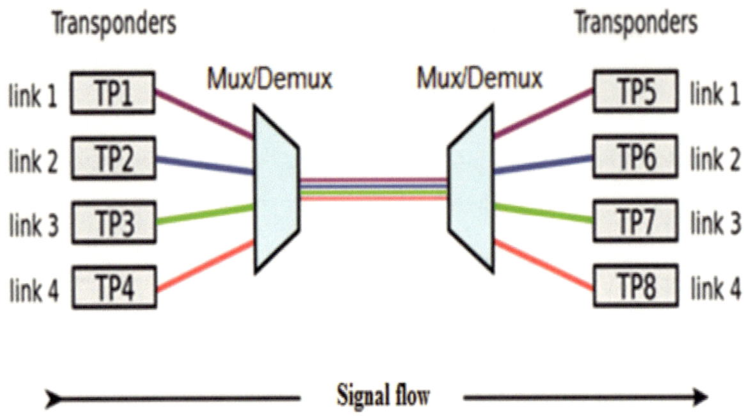

Figure 5.
Bidirectional WDM.

Wavelength division multiplexing is divided into two types, unidirectional WDM and **bi-directional WDM**. In unidirectional WDM, the data is sent only from one side and received on the other side. Multiplexing of the wavelength occurs on the sender side, and demultiplexing of the wavelengths takes place on the receiver side. In bi-directional WDM the data can be sent from both sides which means that both sides can do multiplexing and demultiplexing as shown in **Figure 5**.

2.2.1 Advantages of WDM

- In WDM full-duplex transmission is possible.
- It is easier to reconfigure.
- Optical components are more reliable and provide higher bandwidth.
- Provide high security and faster access to new channel.
- Low cost and easy system expansion.
- Simultaneous transmission of various signals.

2.2.2 Disadvantages of WDM

- Scalability is a concern as optical line termination (OLT); optical line termination has to have transmitter array with one transmitter for each optical network unit (ONU). Adding a new ONU could be a problem unless transmitters were provisioned in advance. Each ONU must have a wavelength-specific laser.
- The cost of the system increases with addition of optical components.
- Inefficiency in BW utilization, difficulty in wavelength tuning, and difficulty in cascaded topology.

2.3 Frequency division multiplexing

FDM [1–4] is based on sharing of the available bandwidth of a communication channel among the signals to be transmitted. It is an analog multiplexing technique that uses a single transmission medium which is divided into several frequency channels. Here the total bandwidth of the channel must be higher than the sum of the individual bandwidth. If the channels are closer to each other, then cross talk may occur; thus, it is necessary to implement channel synchronization. For that purpose some bandwidth is allocated as guard band; these are unused channels placed between the successive transmission channels to avoid cross talk.

For frequency division multiplexing if the input signal is digital, it must be converted to analog before giving it as the input to the modulator.

2.3.1 FDM transmitter block diagram

In FDM signals are generated by sending devices and there are multiple input lines. From the block diagram (**Figure 6**), channel 1 to channel n are taken as the input channels. These signals reach at the input of the corresponding modulator where it receives another signal from a crystal oscillator known as carrier signal, which is a high-frequency high-amplitude signal. The carrier signal is modulated with the input signal. Different modulators use different carrier signals for modulation. It should be noted that the frequency band of one modulator will not make any interference to the frequency band of other modulators.

Each of the modulator produces the corresponding modulated signal at their output. All the output of the modulators will be given to an adder or mixer circuit; from there it is given to another modulator for further shift of total bandwidth. Finally, this higher-frequency signal will be transmitted over the channel.

2.3.2 FDM receiver

The following block diagram (**Figure 7**) shows the concept of demodulation of FDM signal at the receiving side. The antenna receives the multiplexed modulated signal from the transmitter. This signal will be weak at the receiver. Therefore it is necessary to amplify the signal. This is done at the initial stage of the receiver. The amplified signal is then forwarded to the demodulator. The output of the

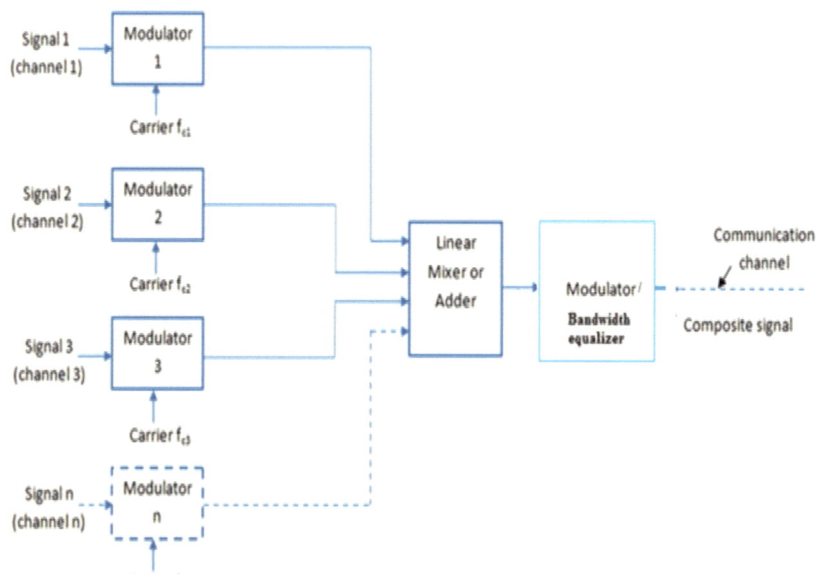

Figure 6.
Block diagram of FDM transmitter.

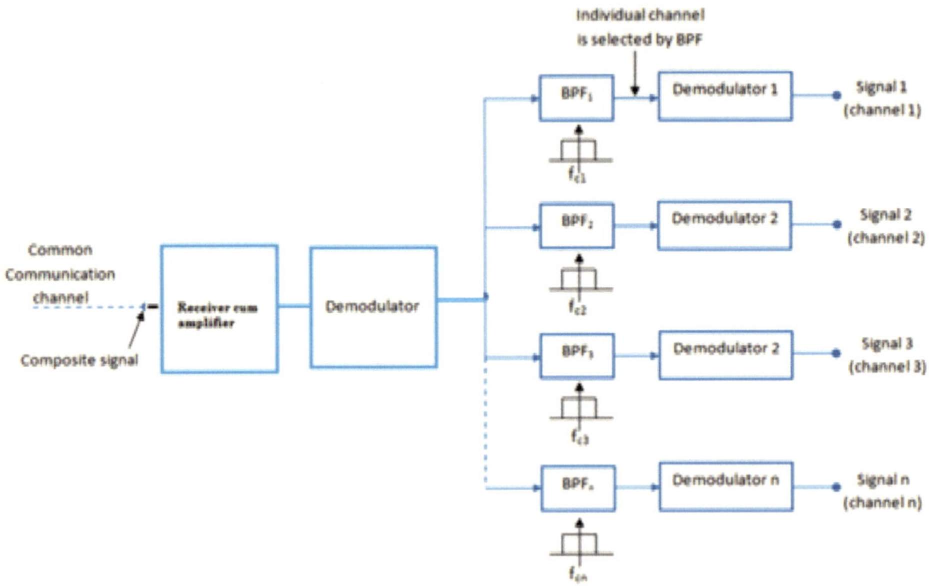

Figure 7.
Block diagram of FDM receiver.

demodulator will be given to the band-pass filters which are well designed with the central frequencies of the carriers as used individually at the transmitting side. Thus the output of each BPF will be the same as that of the originally modulated output of the corresponding modulator.

Then we use the corresponding individual demodulators to recover the original signal.

2.3.3 Advantage of FDM multiplexing

- A large number of signals (channels) can be transmitted simultaneously.
- Demodulation of FDM is easy.
- FDM does not need synchronization between its transmitter and receiver for proper operation.
- Due to slow narrowband fading, only a single channel gets affected.

2.3.4 Disadvantages of FDM

- The communication channel must have a very large bandwidth.
- Intermodulation distortion takes place.
- A large number of modulators and filters are required.
- FDM suffers from the problem of cross talk.
- All the FDM channels get affected due to wideband fading.

2.3.5 Applications of FDM

- FDM is used for FM and AM radio broadcasting.

- FDM is used in television broadcasting.
- First-generation cellular telephone also uses FDM.

2.4 Time division multiplexing

In time division multiplexing (TDM) [1–4], all signals operate with the same frequency at different times, i.e., it is a technique of transmitting several signals over a single communication channel by dividing the time frame into equal slots. Here the signal transmitted can occupy the total bandwidth of the channel, and each signal will be transmitted in its specified time period only. In TDM all signal operates at same frequency at different time slots.

Figure 8 shows the schematic diagram of implementation of TDM system. From this it is clear that a circular ring has been split into eight equal segments and is completely separated from one another. It is also noted that there is a movable arm attached to the inner ring, and it slides over the eight segments over the ring. The eight segments are eight inputs, and the selector moves in clockwise direction from A to H; after completing one revolution, it starts again. The output is taken from the inner ring that contains the signal from only one slot at a time.

The same arrangement is also made at the receiving side. The two inner rings of the transmitting and receiving stations are rotated at the same speed, and they are synchronized. Thus the signal at segment A of the transmitter will reach segment A of the receiver in the period the arm is contacting the segment A. The same is in the case of other segments.

Time division multiplexing is used when data transmission rate of media is greater than the total transmission rate of the sources. Here each signal is allotted to a definite amount of time. These slots are too small so that the transmission appears to be parallel. In TDM all the signals to be transmitted are not transmitted simultaneously. Instead, they are transmitted one by one. When all the signals are transmitted once on the transmission channel, it is said to be one cycle of completion.

Synchronization between the multiplexer and demultiplexer is a major issue in TDM. If the multiplexer and the demultiplexer are not properly synchronized, a bit belonging to one channel may be received by another channel. Therefore, one or more synchronization bits are generally added to the beginning of each frame. These bits, called framing bits, allow the demultiplexer to synchronize with the incoming

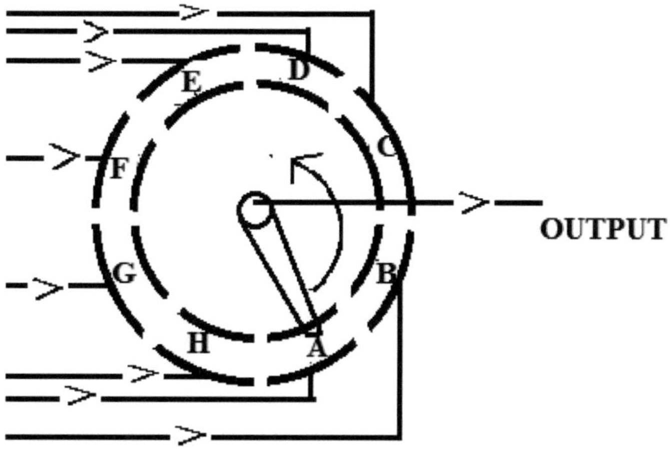

Figure 8.
Schematic diagram of TDM operation.

stream so that it can separate the time slots accurately. Normally, this synchronization information consists of 1 bit per frame, alternating between 0 and 1.

There are two types of TDM multiplexing, synchronous TDM and asynchronous or statistical TDM or intelligent TDM.

2.4.1 Synchronous time division multiplexing

In synchronous TDM the slots are arranged in a round robin manner, i.e., if there are n sources, then a single frame consists of n time slots, and each time slot is dedicated to exactly one source for carrying data from the corresponding input. Each source places its data to the link only when the corresponding slot arrives. In synchronous TDM, if a device does not have data to send, then its time slots remain empty. The transmission of data with synchronous TDM is shown in **Figure 9**.

2.4.1.1 Disadvantages

- The channel capacity cannot be fully utilized when some source do not want to send the data.
- The capacity of the transmission link must be higher than the total capacity of input lines.
- It is very complex to implement.

2.4.2 Asynchronous time division multiplexing

In synchronous TDM if a particular terminal has no data to transmit at a particular time period, the corresponding slot in a frame is wasted or an empty slot will be transmitted. Asynchronous TDM or statistical TDM is used to overcome this difficulty. It dynamically allocates the time slots on the demand to separate input channels, thus saving the channel capacity. Here the time slots are flexible, and the total capacity of input lines can be greater than the link capacity of the channel. In synchronous TDM if there are n input lines, there must be n time slots, but in asynchronous TDM if we have n input lines, then the frame may contain less than n slots. Here the number of slots in a frame is based on a statistical analysis of the number of input lines. The transmission of data with asynchronous TDM is shown in **Figure 10**.

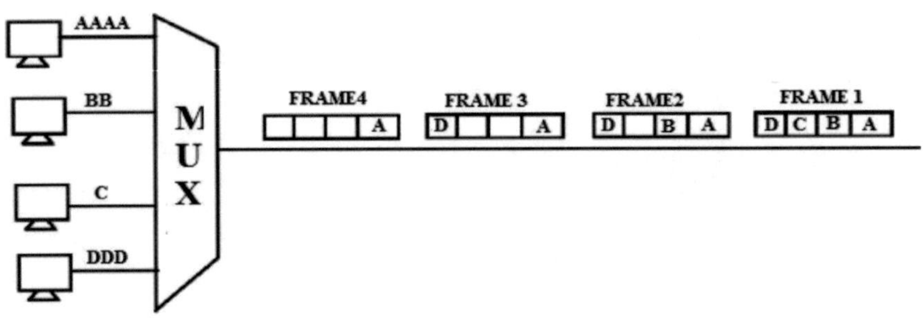

Figure 9.
Synchronous TDM transmission representation.

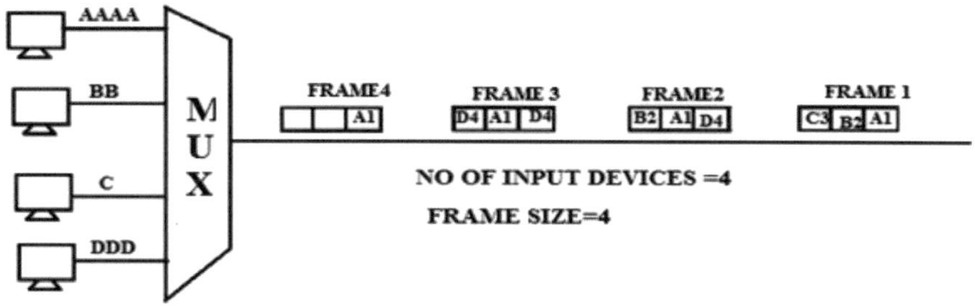

Figure 10.
Data transmission with asynchronous TDM.

2.4.2.1 Disadvantages

- Frames have different sizes.
- An output slot in synchronous TDM is totally occupied by data, in statistical TDM; a slot needs to carry data as well as the address of the destination.
- It requires buffer, and address information is needed as there is no separate slots assigned for each user.

2.5 Code division multiplexing

Code division multiplexing (CDM) [3] is a form of multiplexing in which the transmitter encodes the signal by using a unique chip code which is generated by a pseudorandom sequence generator. It uses spread-spectrum communication, and a narrowband signal is spread over a large band of frequency; it allows multiple signals from multiple users to share a common communication channel. CDM involves the modulation of data with this spreading code in the transmitter side.

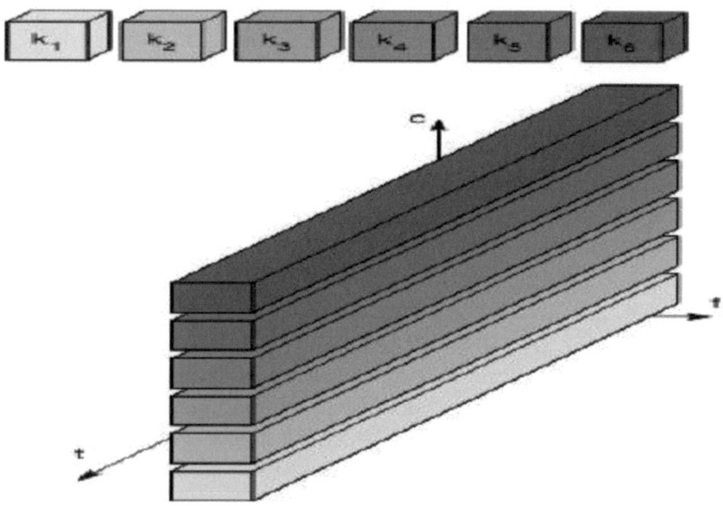

Figure 11.
Code division multiplexing (CDM).

The receiver also wants to know the same code used at the transmitter side in order to decode the signal at the receiving side. Here different random sequences correspond to different communication channels from different stations. To separate other channels, CDM assigns each channel with its own code. The main advantage of CDM is protection from interference and tapping because only the sender and the receiver know the spreading code (**Figure 11**).

CDM is widely used in second-generation and third-generation wireless communication network.

2.5.1 Advantage of CDM

- The CDM does not require any synchronization.
- In CDM more number of users can share the same bandwidth.
- It is scalable.
- It is well-matched with other cellular technologies.
- Interference is reduced because different code words are allocated to each user.
- Efficient utilization of fixed frequency spectrum.

2.5.2 Disadvantages

- The system is more complicated.
- As the number of users increases, the overall quality of services decreases.
- More complex system and primarily it is used in wireless transmission.

2.6 Orthogonal frequency division multiplexing

Orthogonal frequency division multiplexing (OFDM) [4, 5] is a multiplexing technique used in broadband communication system. It is a multicarrier modulation scheme. Now it is used in 4G broadband communication system and next-generation systems. OFDM is popular in broadband wireless systems due to its resistance to multipath fading. OFDM has high data rate capability with reasonable computational complexity. OFDM divides a broadband channel into multiple parallel narrowband subchannels, and each channel carries a low data rate stream of signals. Finally these signals are summing and then transmit as a high data rate stream. In an OFDM transmitter, the input signal bits are mapped into a bank of quadrature amplitude modulator which encodes these into complex symbols. This is fed to an inverse fast Fourier transform (IFFT) to ensure the orthogonality of the subchannels. This output is converted into parallel to serial, modulated into a carrier wave, and then transmitted into the air. At the receiver the reverse process is performed for recovering the original signal. The advantages of OFDM are that its low computational complexity because OFDM may be viewed as a many slowly modulated narrowband signals rather than a rapidly modulated wideband signal.

2.6.1 OFDM transmitter

The block diagram depicting the OFDM transmitter is shown in **Figure 12**. For OFDM transmitter, a serial stream of binary digits is considered as the input. The input is converted into N parallel streams using inverse multiplexing. The transformation of N parallel streams into the state-space mechanism is performed by means of modulation techniques like quadrature amplitude modulation (QAM) and phase shift keying (PSK).

A digital modulation system for data communication by varying or modulating the phase of the reference signal or the carrier wave signal is known as PSK. A finite number of phases is involved in PSK with each phase having a distinctive pattern of binary digits. An integration of trouble-free AM and simple phase modulation is called QAM in which the large amount of data is transmitted over the same bandwidth due to the synergistic effect of simple amplitude modulation and phase modulation. Hence, QAM increases the efficiency of data transmission for radio communication systems (**Figure 13**).

2.6.2 Block diagram of OFDM transmitter

To provide a set of complex time-domain samples, IFFT is calculated for each set of symbols. Later, the time-domain samples are quadrature mixed to passband in the normal way. By the use of digital-to-analog converters (DACs), the real and imaginary components are primarily converted to the analog domain. Such analog signal helps to modulate corresponding cosine and sine waves at the carrier frequency. Finally, those signals are summed up to provide the transmission signal.

2.6.3 OFDM receiver

At the OFDM receiver's side, the reverse process of transmitter side is performed. The block diagram representing the OFDM receiver is shown in **Figure 14**. The transmitter-generated signal is further transmitted over the channel for receiving. The receiver receives the baseband OFDM signals, and then it passes through a low-pass filter to remove the unwanted signals. The baseband signals are then sampled and digitized using ADCs, and a forward FFT is used to convert back to the frequency domain.

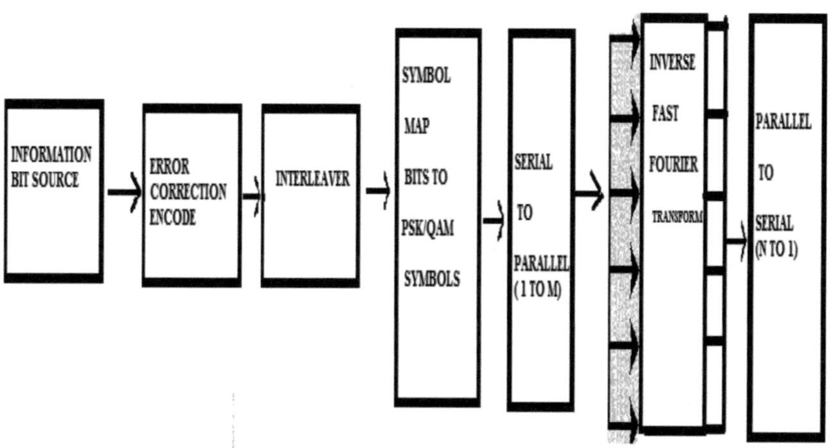

Figure 12.
OFDM transmitter.

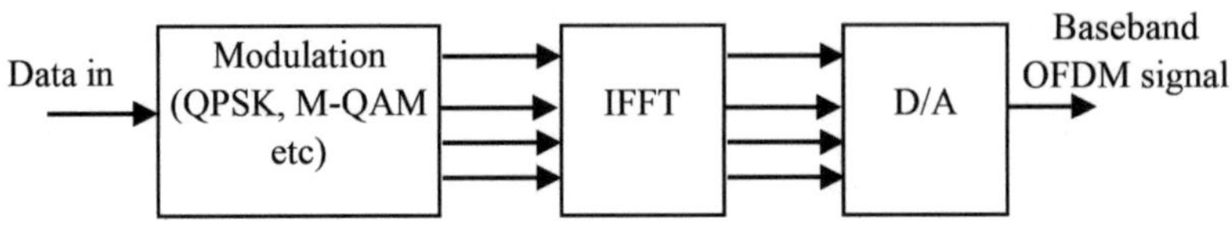

Figure 13.
OFDM transmitter simple block diagram.

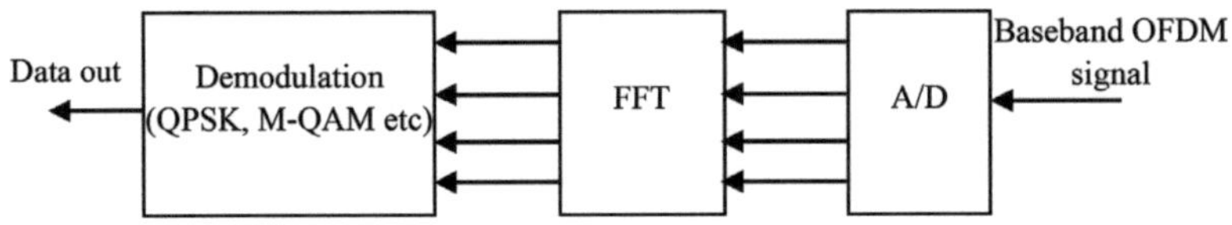

Figure 14.
OFDM simple block diagram of OFDM receiver.

By means of an appropriate symbol detector, the frequency domain signals are converted to N parallel streams, and each stream is converted to a binary stream. A sequential stream combining all binary stream acts as an estimate of the original binary stream at the transmitter side.

2.6.4 Significance of the OFDM system

- OFDM is computationally efficient to deploy the modulation and demodulation techniques through IFFT and FFT, respectively.

- OFDM signal is robust and more tolerant in multipath propagation environment to delay spread.

- OFDM is more resistant to frequency selective fading than single carrier transmission systems.

- OFDM system gives good protection against co-channel interference and impulsive parasitic noise.

- OFDM system uses pilot subcarriers to prevent the frequency and phase shift errors.

OFDM system has also certain limitations rather than the abovementioned potential capabilities. High peak-to-average power ratios (PAPR) of transmitted signal are the major drawback of the OFDM signal. OFDM is very sensitive to carrier frequency offset and hence becomes difficult to synchronize during sharing of subcarriers different transmitters.

3. Digital multiplexer

Digital multiplexer [6–8] or data selector is a logic circuit that has several input lines and a single output line. It also consists of data selector switch which is used to select the inputs and permit the data into the device to output.

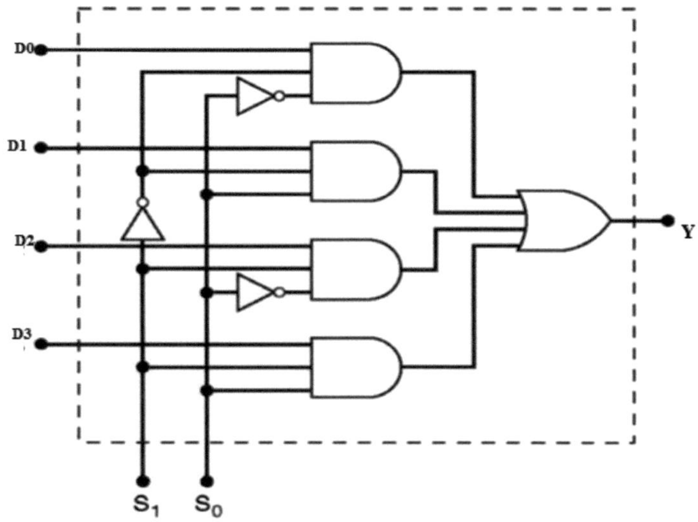

Figure 15.
Circuit diagram of four-in-one multiplexer.

SYMBOL OF 4 IN 1 MUX

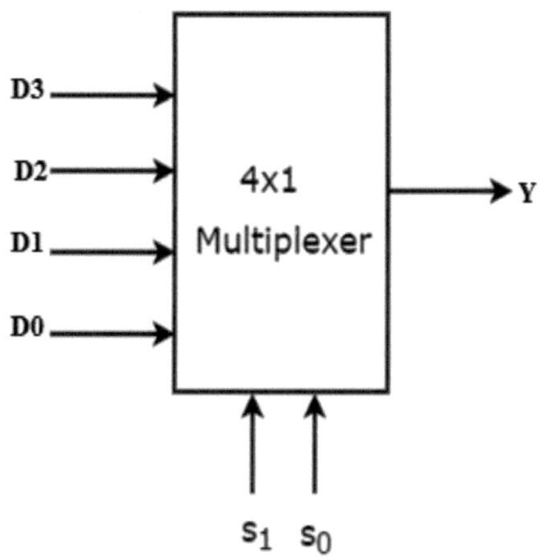

SELECTOR SWITCH

S1	S0	INPUT SELECTED
0	0	**D0**
0	1	**D1**
1	0	**D3**
1	1	**D4**

Figure 16.
Circuit symbol and selector switch pattern of four-in-one multiplexer.

SYMBOL OF 8 IN 1 MUX

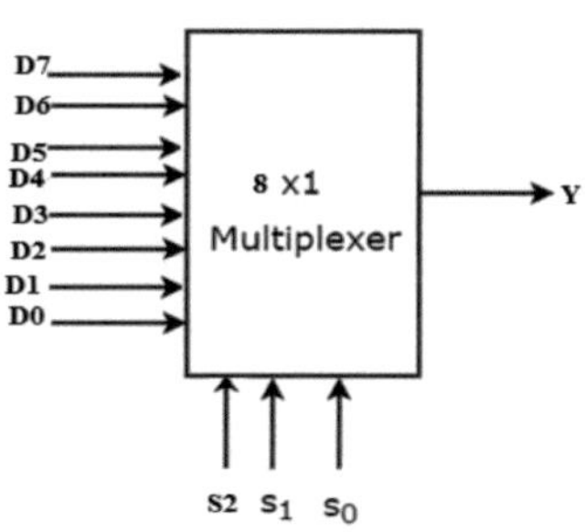

DATA SELECTOR

S2	S1	S0	INPUT SELECTED
0	0	0	D0
0	0	1	D1
0	1	0	D2
0	1	1	D3
1	0	0	D4
1	0	1	D5
1	1	0	D6
1	1	1	D7

Figure 17.
Logic symbol and switching pattern of eight-in-one multiplexer.

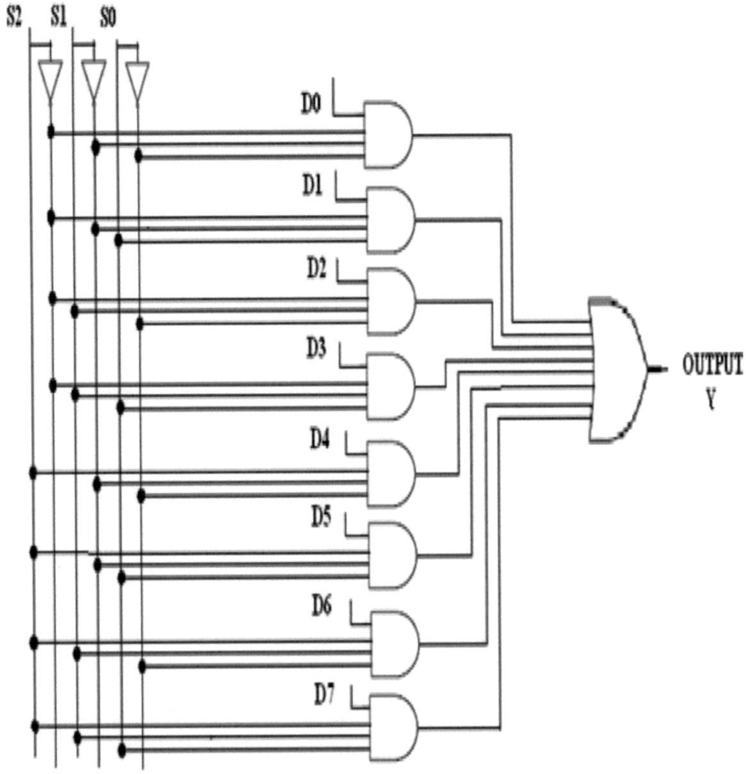

$$Y = E.[\overline{S2}\,\overline{S1}\,\overline{S0}\,D0 + \overline{S2}\,\overline{S1}\,S0\,D1 + \overline{S2}\,S1\,\overline{S0}\,D2 + \overline{S2}\,S1\,S0\,D3 + S2\,\overline{S1}\,\overline{S0}\,D4 + S2\,\overline{S1}\,S0\,D5 + S2\,S1\,\overline{S0}\,D6 + S2\,S1\,S0\,D7]$$

Figure 18.
Logic diagram of eight-in-one multiplexer.

3.1 Four-in-one multiplexer

The logic symbol and circuit for a four-input multiplexer are shown in **Figure 15**.

Here D0, D1, D2, and D3 are data input lines. S0 and S1 are data selector or logic switches. When S0 = S1 = 0, then the two inputs of the first AND gate become actively high because the selectors S0 and S1 are inverted using NOT gate and given to this gate. Thus the data from D0 line is outputted through this AND gate. At that time the other gates are in 0 output position. Similarly when S0 = 1 and S1 = 0, then the two inputs of the AND gate 2 become actively high; thus, data from D1 is transmitted through gate 2 as output, and all other gates are in 0 output position. In this manner D2 and D3 are inputted to consecutive switch positions. Here an OR gate is used to combine these four output lines as a single output (**Figure 16**).

3.2 Eight-in-one multiplexer

The logic symbol and data selector of eight-in-one multiplexer is shown in **Figure 17**.

If input is 2, then one data selector switch is needed; if input is 4, then two selector switches are needed; if input is 8, then three selector switches are needed; if input is 16, then four selector switches are needed; and so on.

In an eight-in-one multiplexer (**Figure 18**), eight data input lines such as D0, D1, D2, D3, D4, D5, D6, and D7, data selector S0, S1, and S2. In this circuit these inputs are fed to eight AND gates, and the outputs of the AND gates are combined by using an OR gate.

3.2.1 Working

When the three selector switches are actively low, then the three inputs of the first AND gate become actively high because the selector outputs are NOTed and given to the first AND gate. Thus the data from D0 line is outputted through the first AND gate, and all other AND gates are in 0 output position. When S0 = 1, S1 = 0, and S2 = 0, then the three input terminals of the second AND gate become actively high, and D1 is outputted through this gate. At that time all other gates are in 0 output position. In that manner D2, D3, D4, D5, D6, and D7 data are outputted in the next consecutive switch positions.

Important TTL multiplexer IC's are as follows:

74,157—2 input data selector/MUX.

74,151—8 input data selector/MUX.

74,150—16 input data selector/MUX.

3.2.2 Application of MUX

- Seven-segment display unit
- Function generators
- Digital counters
- Parallel-to-serial conversion

4. Demultiplexers

Demultiplexer [6] is a logic circuit that performs the reverse multiplexer function. Demultiplexer receives signal from a single line (serial input) and transmits these information into multiple output lines and parallel output lines (**Figure 19**).

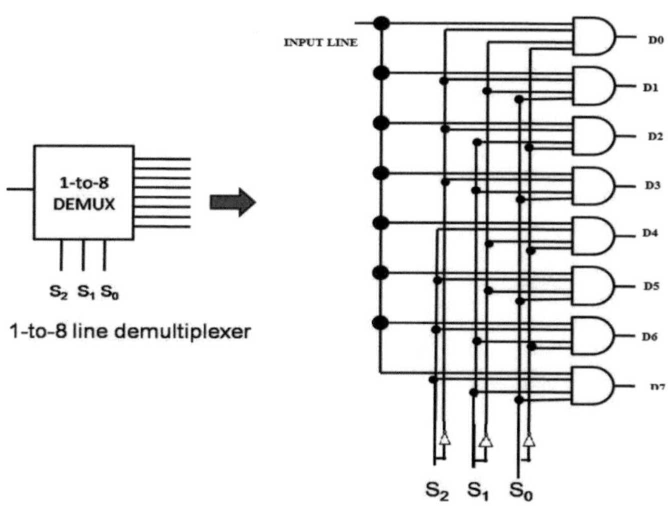

Figure 19.
Circuit diagram of one-to-eight demultiplexer.

5. Major digital modulation schemes

In analog communication we use different modulation schemes like amplitude modulation, frequency modulation, phase modulation, etc. [9]. In digital communication we use the following schemes:

- Amplitude shift keying (ASK)
- Frequency shift keying (FSK)
- Phase shift keying (PSK)

5.1 Amplitude shift keying

The principle of amplitude shift keying is that the amplitude of the carrier wave is modulated in accordance with the digital message signal, i.e., ASK signal represents the binary data in the form of variations in the amplitude of the carrier signal. When an ON condition of digital pulse exists, then carrier will be switched ON, and when an OFF condition encounters, the carrier will be switched OFF. The time period for which the carrier is present or absent depends on the time interval for which the unipolar pulses are present.

Here the amplitude of carrier signal is varied to represent binary 1 and binary 0 data inputs, while the frequency and phase of the carrier signal remain constant. Voltage levels are left to designers of the modulation system (**Figure 20**).

5.1.1 ASK advantages and disadvantages

The major advantage of ASK includes high bandwidth efficiency and simplicity in its design. In ASK the modulation and demodulation processes are comparatively inexpensive.

Its disadvantages include lower power efficiency, and it is very susceptible to noise interference.

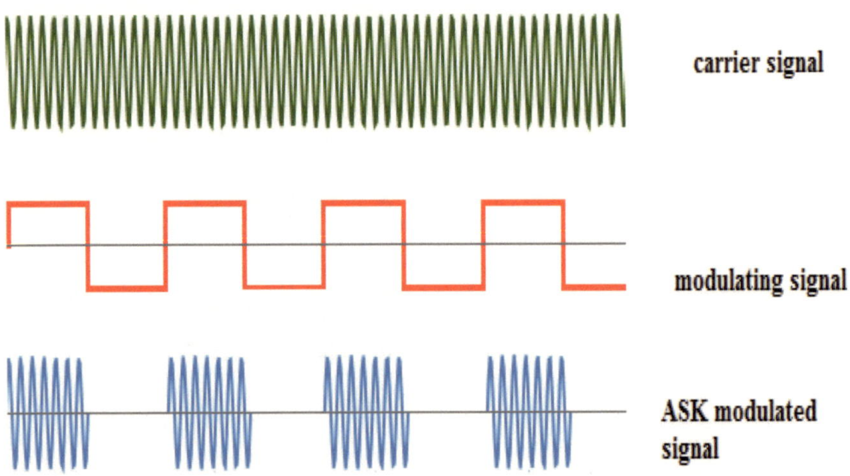

Figure 20.
Amplitude shift keying.

5.1.2 Application

- Used in our infrared remote controls
- Used in fiber optical transmitter and receiver

5.2 Frequency shift keying

The principle of frequency shift keying is that the frequency of the carrier wave is modulated in accordance with the digital message signal, i.e., FSK signal represents the binary data in the form of variations in the frequency of the carrier signal. When an ON condition of digital pulse exists, then carrier will be switched to one frequency, and when an OFF condition encounters the carrier, it will be switched to another frequency, i.e., in "frequency shift keying," the frequency of a sinusoidal carrier is shifted between two discrete values (**Figure 21**).

5.2.1 Advantages and disadvantages of FSK

Its advantages include lower probability of error and provide high signal-to-noise ratio. It has higher immunity to noise due to constant envelope. Therefore the probability of error-free reception of data is high. FSK transmitter and FSK receiver implementations are simple for low data rate application.

The disadvantage of FSK includes the usage of larger bandwidth than other modulation techniques such as ASK and PSK. Hence it is not bandwidth efficient and is extensively used in low-speed modems having bit rates below 1200 bits/sec. It is not preferred for the high-speed modems because with increase in speed, the bit rate increases.

5.2.2 Application

Many modems used FSK in telemetry systems.

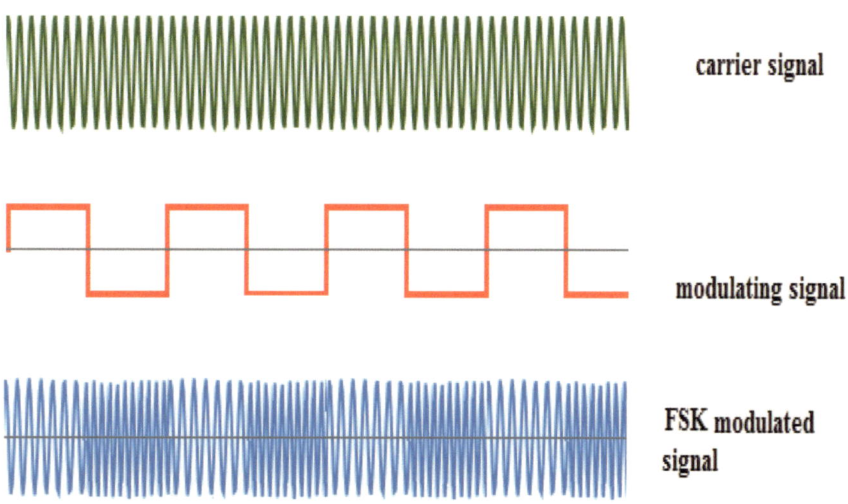

Figure 21.
Frequency shift keying.

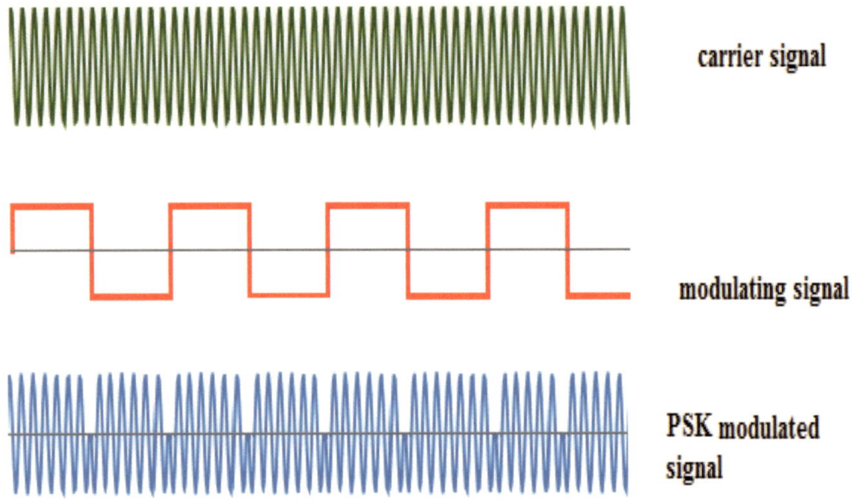

Figure 22.
Phase shift keying.

5.3 Phase shift keying

The principle of phase shift keying is that the phase of the carrier wave is modulated in accordance with the digital message signal, i.e., PSK signal represents the binary data in the form of variations in the phase of the carrier signal. Here the carrier signal changes its phase depending on the nature of the transmitted bit, i.e., 0 or 1. In two levels of PSK, the difference of 180-phase shift is used between binary 1 and binary 0 (**Figure 22**).

5.3.1 Advantages and disadvantages of PSK

It is a more power-efficient modulation technique than the ASK and FSK as it is less susceptible to errors than ASK modulation and occupies the same bandwidth as ASK. The disadvantage of PSK includes lower bandwidth efficiency. The binary data is decoded by estimation of phase states of the signal. These detection and recovery algorithms are very complex. It is also one form of FSK, and hence it also offers lower bandwidth efficiency than ASK modulation type.

5.3.2 Application

1. Used in our ADSL broadband modem

2. Used in satellite communication

3. Used in our mobile phones

6. Summary

This chapter "multiplexing" deals with different multiplexing techniques commonly used in both analog and digital communication systems and their classifications. This chapter contains the detailed description of analog modulation techniques like FDM, WDM, etc. It also describes about OFDM techniques and various digital modulation schemes like TDM and its variants. It also focuses on

DOI: http://dx.doi.org/10.5772/intechopen.85866

CDM multiplexing techniques and different digital multiplexers and demultiplexers and varies in digital demodulation techniques like ASK, PSK, and FSK with detailed diagrams.

Author details

Vijayakumar Nandalal* and M.S. Sumalatha
Department of Electronics and Communication Engineering, Sri Krishna College of Engineering and Technology, Coimbatore, Tamilnadu, India

*Address all correspondence to: nandalalskcet@gmail.com

References

[1] Stallings W. Data and Computer Communications tenth edition. Upper Saddle River: Pearson Education, Inc. Pearson Prentice Hall Pearson Education, Inc; 13 September 2013

[2] Forouzan BA. Data Communications and Networking. 4th edition. McGraw Hill Higher Education

[3] Schiller J. Mobile Communications. tenth edition 13 September 2013 Pearson Education, Inc. Pearson Prentice Hall Pearson Education, Inc. Upper Saddle River, NJ 07458 Second Edition

[4] Schiller JH. Mobile Communications 2nd edition Pearson Education India. 2008;**2**. Available from: https://nptel.ac.in/courses/106105080/pdf/M2L7.pdf

[5] Pechetty TR, Vemulapalli M. An implementation of OFDM transmitter and receiver on reconfigurable platforms. International Journal of Advanced Research in Electrical, Electronics and Instrumentation Engineering. November 2013;**2**(11):5486-5490

[6] Floyd TL. Digital Fundamentals. 11 th Edition 2015. Upper Saddle River, NJ, USA: Pearson publication Prentice-Hall Inc; ISBN: 9780132737968, 0132737965. 2013

[7] Nair BS. Digital Electronics and Logic Design. PHI Learning Private Limited. 1 January 2002; ISBN: 9788120319561

[8] Yarbrough JM. Digital Logic Application and Design. 1 edition Cengage Learning; ISBN-10: 9788131500583. 2006

[9] Available from: https://www.tutorialspoint.com/principles_of_communication/principles_of_communication_digital_modulation_techniques.htm

Chapter 3

Overview of Multiplexing Techniques in Wireless Networks

Pedram Kheirkhah Sangdeh and Huacheng Zeng

Abstract

Multiplexing is a key technique for wireless communications. In this chapter, we will review the multiplexing techniques that are used in exiting wireless systems.

Keywords: wireless networks, multiplexing, CDMA, TDMA, FDMA, NOMA

1. Introduction

With the great advances of communication technologies, the communication paradigm has widely been shifted from point-to-point to multi-user wireless systems to support ever-increasing number of mobile devices being introduced in the market. The proliferation of mobile devices has necessitated an elaborate mechanism to serve multiple users over a shared communication medium. The most important building block in this mechanism is multiplexing approach. The multiplexing refers to a method which aims at combining multiple signals into one signal such that each user would be able to extract its desired data upon receiving the multiplexed signal. **Figure 1** shows a communication system with three sources and corresponding destinations at system level. As shown in **Figure 1(a)**, the system without multiplexing requires three different communication links each of which carrying the signal of single source toward its destination exclusively. Such a system is inefficient since it demands triple times more communication resources than the same system with multiplexer/de-multiplexer shown in **Figure 1(b)**. With the aid of multiplexing, signals of all sources will be superimposed into one signal and sent over single available communication link. Establishing a successful transmission over the single link preserves valuable resources and decreases communication costs. Furthermore, serving multiple users through a channel result in massive connectivity, which paves the way for current and next generation of wireless networks designed for crowded urban areas. In today's communication, multiplexing has penetrated many communication applications ranging from digital broadcasting to Wi-Fi networks. Multiplexing brings the following advantages to wireless communication systems.

- Reducing the communication cost.
- Enhancing the connectivity.
- Saving valuable communication resources.

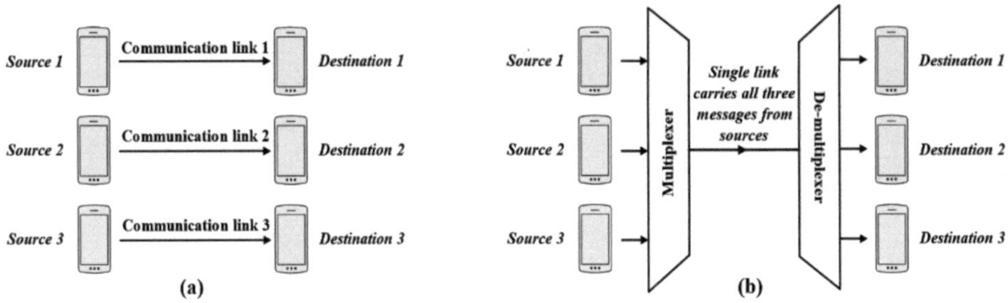

Figure 1.
A communication system comprising three sources and destinations: (a) without multiplexing and (b) with multiplexing.

- Improving network capacity.

- Eliminating the need of exclusive links between sources and destinations.

Based on the application requirements, available spectrum, and users' hardware capability, the appropriate multiplexing approach will be designed over the required domain which could be time, frequency, power, code, wavelength or delay-Doppler domain. For example, time division multiplexing is not suitable for a delay-sensitive application, and spatial multiplexing is very amenable for a network with multi-antenna nodes. Among all of the aforementioned domains for multiplexing, only wavelength division multiplexing exclusively targets communication over the fiber cables while others suit wireless communication. In this chapter, the most important multiplexing approaches are studied.

2. Time division multiplexing

Time division multiplexing (TDM) is the first multiplexing scheme has been introduced to be employed in wired and wireless networks. Generally speaking, TDM separates signals from different sources over non-overlapping time slots to share the same spectral medium. At the receiver side, upon detecting the time slots, the desired signals can be recovered through overhearing the related slots. The simplest TDM can be modeled as a switch that periodically moves between multiple sources, and the transition time equals to the slot allocated to a single source. The messages of all sources are combined into a frame and sent over the medium. Based on flexibility of the allocation mechanism, TDM can be implemented in synchronous or asynchronous mode. TDM and its variants have been embedded in plethora of recent technologies, such as WiMAX [1], Tactical data link (TDL) [2], Bluetooth [3], HIPERLAN [4], to name but a few.

2.1 Synchronous time division multiplexing

Synchronous TDM follows a strict approach in time slot allocation. It periodically assigns subsequent time slots to different sources in a predefined order, no matter if some sources have nothing to be sent. As shown in **Figure 2**, the multiplexer and sources can be modeled as a switch and states, respectively. The switch circularly moves between different states and remains in different states for equal amount of time, i.e., a time slot. In a cycle, the switch passes through all states and shares the cycle time between all sources fairly. For the sake of simplicity, the transition time of switching is ignored.

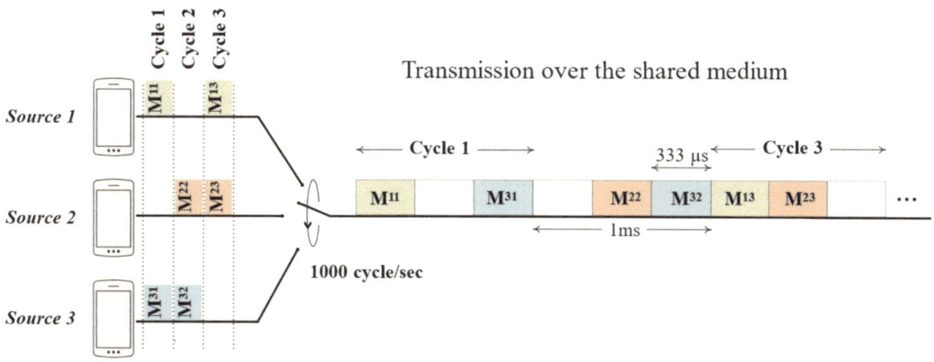

Figure 2.
Synchronous time division multiplexing.

Figure 2 depicts an example of synchronous TDM for three independent sources in three cycles. The switch rotates between states (sources) with the rate of 1000 cycle per second. Hence, the cycle time is 1 ms, and each time slot equals to 333.3 μs. In the jth cycle, the ith source may have message M_{ij} or nothing to be transmitted. For instance, in second cycle, source one remains idle while source two and source three have M_{22} and M_{32} for transmission, respectively. Therefore, these messages occupy the second and third time slots in the second cycle while the second time slot will be wasted without conveying any message. An unused time slot is shown with hatched rectangular. Each cycle could be preceded/terminated with a preamble/postamble enabling destinations to detect beginning/end of a cycle. The detail of each cycle is shown in the figure. Obviously, in each cycle, one third of the airtime will be wasted which drastically degrades the throughput of the system. To prevent squandering airtime, asynchronous TDM has emerged.

2.2 Asynchronous time division multiplexing

Asynchronous TDM, also known as statistical TDM, pursues a more dynamic approach by giving the airtime to sources that have data for transmission. In this manner, the messages of different sources occupy all subsequent time slots, which yield improvement of spectrum utilization in turn. As a well-known application, asynchronous TDM is used in asynchronous transfer mode (ATM) networks [5]. To demonstrate the possible gain of the asynchronous TDM over synchronous TDM, let us consider the sources previously shown in **Figure 2**. This time, the asynchronous TDM is applied on the system, as shown in **Figure 3**. In each cycle, the switch passes through all sources and transfers the existing messages to frame assembler. The frame assembler tags a preamble to each message. The preambles include an ID or address field to notify the origin or intended destination of the message attached to the preamble. Then, the frame assembler aggregates the tagged messages and disregards the idle time intervals related to silent sources. The tagged messages occupy subsequent time slots and are transmitted sequentially. As shown in the figure, each cycle carries messages of sources that have something to transmit; hence, the cycles include two-time slots which improves the spectral utilization by 33%. Although the cycles duration are equal in the example, the transmission duration may vary depending on the number of messages to be carried. Clearly, asynchronous TDM demands more processing capability at multiplexer and de-multiplexer, and it may cause a delay up to one cycle for buffering and aggregating the existing messages. However, it is worthwhile since it yields higher spectral efficiency and saving valuable resources. As another advantage,

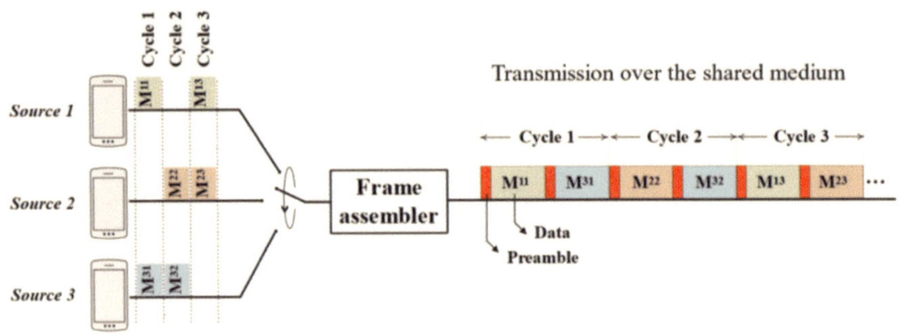

Figure 3.
An example of asynchronous/statistical time division multiplexing.

using the preambles diminishes the need of synchronized clocks between multiplexer and de-multiplexer.

3. Frequency division multiplexing

Frequency division multiplexing (FDM) is a multiplexing technique which divides the available bandwidth into multiple sub-bands each of which is able to carry a signal. Therefore, FDM enables concurrent transmissions over a shared communication medium. As another common use, FDM enables the system to send a huge amount of data through several segments transmitted over independent frequency sub-bands.

Figure 4 reveals basic principles of the conventional FDM. In the figure, signal of three independent sources are multiplexed to be sent over the medium. Each source has a flat signal with large width in time domain. Depending the value of the information bit, the signals can be $x(t) = \pm 1$. Hence, the spectrum of each signal is approximated by $X(f) \approx \pm\delta(f)$. The baseband signals are converted to well-separated bandpass signals by multiplexer. The signals are conveyed by carriers with different frequencies such that each signal is located within a non-overlapping sub-band and does not leak onto other signals. The carrier frequencies spacing, and sub-band width are application-specific and depend on the available bandwidth. The multiplexed signal will be transmitted over the medium. At the receiver side, de-multiplexer employs proper filters to extract the desired bandpass signals. The intended bandpass signals will be converted to baseband signals for further processes at destinations.

The spectrum of a communication system with N sources benefiting from FDM is depicted in **Figure 5**. The multiplexer uses N equally spaced carrier frequencies. The carrier frequency for the ith user is f_{ci}. For further simplicity, consider that

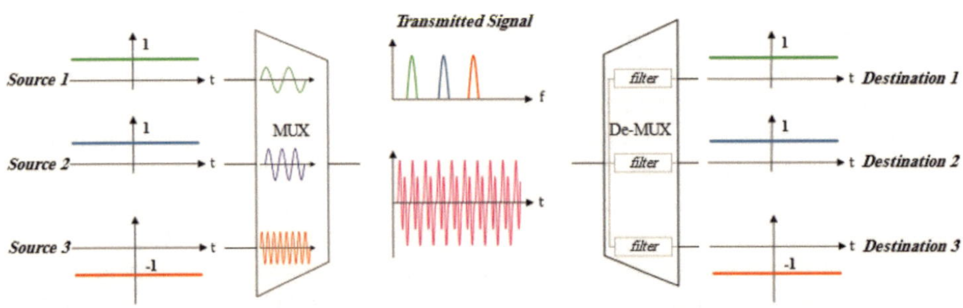

Figure 4.
Frequency division multiplexing.

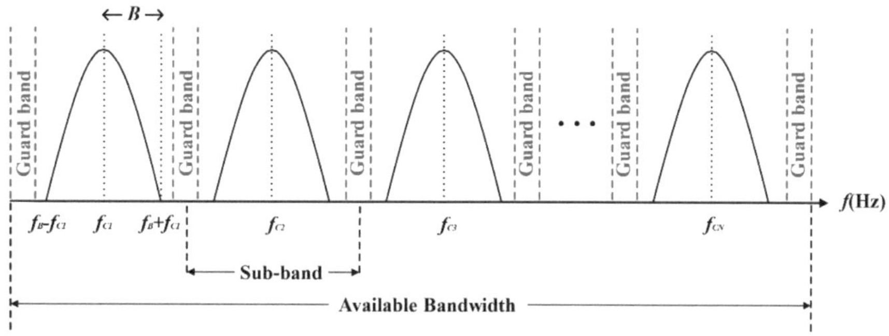

Figure 5.
Spectrum of multiplexed signals via FDM for a system with N *sources.*

sources use the same pulses for data transmission. Signals from the ith source lies within the ith frequency sub-band centered around f_{ci}. Bandwidth of the base-band signals is $B = f_B\ Hz$ which is less than sub-bands' width, i.e., $SB = f_{c(i+1)} - f_{ci}$. To prevent leaking signals into adjacent sub-bands, the sub-bands are separated by a guard band $G = f_G\ Hz$. Clearly, for appropriate signal transmission, the carrier frequency spacing should be designed such that $f_{ci} + f_B < f_{c(i+1)} - f_B$.

3.1 Orthogonal frequency division multiplexing (OFDM)

Conventional FDM allocates the available spectrum to the sources very generously, and the spectral efficiency is not among its concerns. With this motivation, OFDM has been introduced for efficient use of spectrum. OFDM is a multi-carrier modulation through which a data stream, like voice, video, or data, is distributed among multiple subcarriers separated closely and precisely. In a simple OFDM-based system, the modulated samples over frequency domain, i.e., $x[k]$, are distributed over different subcarriers via Inverse Fast Fourier Transform (IFFT) as follows.

$$x(t) = \sum_{k=-\frac{N}{2}}^{\frac{N}{2}-1} x[k] e^{\frac{j2\pi kt}{N}} \tag{1}$$

where $x(t)$ denotes the time domain signal which is sum of multiple sinusoids and $x[k]$ is the kth modulated sample. Since the basis of the transformation is unit vectors with equally angular separated in polar plane, the spectrum of OFDM signal is composed of N shifted *sinc* functions. **Figure 6** illustrates the spectral basis of an OFDM signal with four orthogonal subcarriers. The subcarrier spacing, i.e., $\Delta f = f_{i+1} - f_i$, is chosen such that the center frequency of each subcarrier is located on a null point of other subcarriers. Therefore, at the moment of sampling from the center of subcarriers, no interference is experienced from others. The time domain signal goes through another process called cyclic prefix (CP) insertion to be immune against multipath fading. In this step, a certain amount of the time-domain signal's tail is copied and attached to the beginning of the time-domain signal. If the maximum delay in multipath environment does not exceed the time duration CP, the signal can be recovered perfectly. The CP-inserted OFDM signal is fed into a D/A converter. The baseband analog signal is then converted to a bandpass signal to be sent over the communication medium.

At the receiver side, the superimposition of the analog signals received from different paths within reception interval is sampled and converted to a baseband

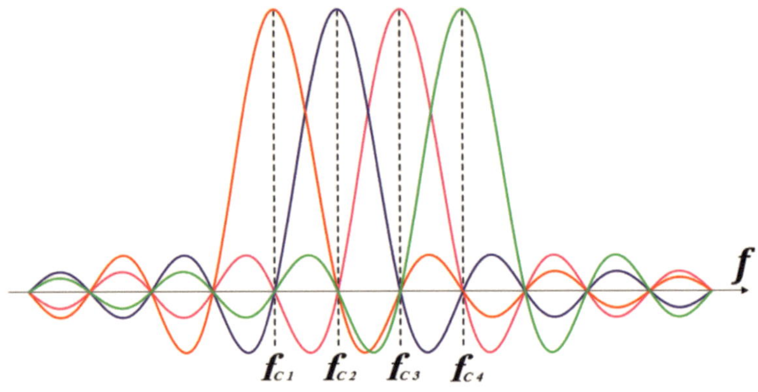

Figure 6.
Bandwidth requirement for an OFDM system using four orthogonally spaced subcarriers.

digital signal. The parallel to serial unit maps the sequence of received OFDM samples into *N* parallel samples. Then, the CP added by transmitter is removed. The remaining samples go through Fast Fourier Transform (FFT). The resulting series of modulated samples is fed into demodulator, and the desired data is recovered.

Due to promising performance of the OFMD technique and its very low computational complexity, this technique has been embedded in many wireless communication systems like IEEE WLAN standards [6–8], LTE/LTE-A [9–11]. The popularity of this technique stems from following advantages which make it highly amenable for real world application.

- *Resilience in frequency selective environments:* OFDM decomposes whole available spectrum to several narrow channel in frequency domain. It is very likely that signals carried over a subcarrier experience a relatively flat channel although the channel may be frequency selective.

- *Resilience to inter-symbol interference (ISI)*: single-carrier communication is vulnerable to ISI specially when the data rate grows. OFDM tackle this problem with dispatching signals over multiple sub-channels. Indeed, OFDM technique changes a transmission with high rate into multiple low-rate transmissions. In this manner, it increases the symbol duration and push the duration beyond maximum delay of the channels. Lack of ISI also means simpler equalization mechanism and reduction in hardware cost of the OFDM receiver. At the end of this section, example 1 reveals how OFDM technique combats the frequency selectivity of the channel.

- *Resilience to narrow-band interference*: narrow-band interference drastically diminishes the throughput of single-carrier systems either by blurring the reference signals for synchronization or corrupting the data. However, if the signal is transmitted using OFDM, only a portion of symbols is contaminated by interference. The erroneous parts caused by interference can be recovered with the aid of error correction codes and interleaving to isolate errors.

- *Spectral efficiency*: comparing **Figures 5** and **6**, it is clear that closely separated frequency sub-channels yields higher spectral efficiency for OFDM.

- *Low-computational complexity*: although OFDM technique is more complex than conventional FDM, it intrinsically demands low-computational capability since it possesses simple mathematical operations. It can be simply implanted with FFT and IFFT modules containing nothing more than adders, multipliers, and registers.

Example 1: Consider a transmission which requires 1 Mbps data rate and RMS delay spread is 10 μs and frequency selectivity condition is $\tau_{rms} > T_{sym}/10$ where T_{sym} denotes symbol duration. The comparison of single-single carrier transmission and OFDM technique with 128 subcarriers is as follows:

- *Single-carrier case*: $T_{sym} = 1\ \mu s$ and $\tau_{rms} > 1\ \mu s/10$. Therefore, ISI is imminent.

- *Multi-carrier case*: in this case, data rate of a subcarrier is 7.8125 kbps. The T_{sym} for a subcarrier equals to 128 μs. Then, $\tau_{rms} < 128\ \mu s/10$, and the signal carried by each subcarrier experiences a flat fading channel.

In spite of aforementioned advantages, the OFDM technique suffers from two main disadvantages. High peak-to-average power ratio (PAPR) stems from the large range of amplitude of the OFDM signal and impedes proper performance of amplifiers at OFDM transceivers. Also, OFDM technique is very sensitive to carrier frequency offset (CFO). This effect emerges due to hardware impairment between transmitter and receiver local oscillators and cause inter-carrier interference (ICI).

4. Code division multiplexing

Code division multiplexing (CDM), also called spread-spectrum technique, is a multiplexing technique which has been widely implemented in third generation of wireless network. It takes full advantage of the available spectrum. Through several concurrent transmissions over the spectrum, this technique has enhanced the capacity of network 18 times compared to first generation and 6 times compared to the second generation of wireless communication technologies. For transmitting multiple messages over the channel simultaneously, the multiplexer assigns a separate spreading code from a set of orthogonal pseudo-random sequences to each user. The orthogonality of these sequences will help users to recover their desired signals from the multiplexed signals.

Consider a system with 4 users as depicted in **Figure 7**. AP intends to send one bit to each user, say b_i for user i. The sequence c_i is used to encode/decode the bits

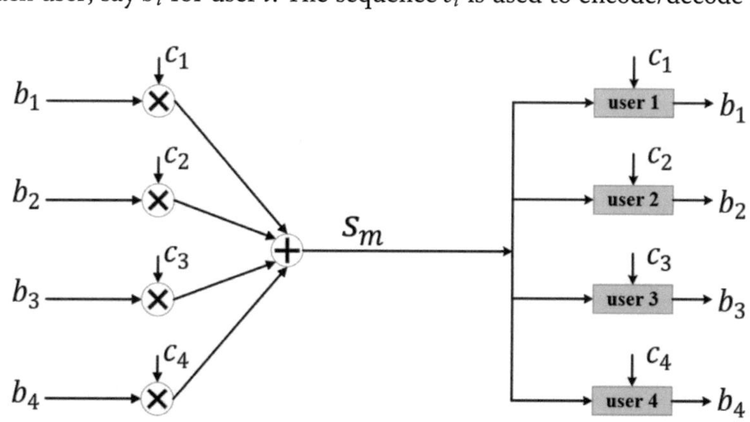

Figure 7.
An example of CDM for a network with four users.

to/from user i. These sequences are devised such that the inner product of two different sequences is zero, i.e., $c_i.c_j=0\ i\neq j$. The inner product of a sequence with itself is M which is the number of users. Also, the '0' is mapped to −1, and 1 is mapped to +1. The multiplexed sequence is $s_m=\sum_{i=1}^{4} b_i c_i$. Upon receiving the multiplexed sequence, user $\boldsymbol{i}$ recovers its desired information by multiplying its corresponding spreading sequence and received sequence and dividing the result by the length of sequences. The key component in CDM is the spreading code. Walsh code is among the most popular sequences used for CDM. Walsh codes with length of 2^n, $n \in \mathbb{N}$, can be constructed with Hadamard matrices as follows.

$$H_1 = [1] \rightarrow H_2 = \begin{bmatrix} 1 & 1 \\ 1 & -1 \end{bmatrix} \rightarrow \ldots \rightarrow H_{2N} = \begin{bmatrix} H_N & H_N \\ H_N & -H_N \end{bmatrix} \quad (2)$$

where each row of H_{2N} can be used as a sequence with length $2N$. This procedure results four orthogonal sequences as follows.

$$c_1 = [1, 1, 1, 1], c_2 = [1, -1, 1, -1], c_3 = [1, 1, -1, -1], c_4 = [1, -1, -1, 1] \quad (3)$$

Consider the following set of bits as an example, $b_1=b_2=b_3=1$, and $b_4=-1$. Therefore, the multiplexed signal is:

$$s_m = \sum_{i=1}^{4} b_i c_i = [+2, +2, +2, -2]. \quad (4)$$

The first user recovers its desired data $\hat{b}_1=s_m.c_1^T/4$. And other users do the same procedure with their own sequences.

$$\hat{b}_1 = \frac{1}{4}[1, 1, 1, 1] \times [2, 2, 2 - 2]^T = \frac{4}{4} = +1. \quad (5)$$

Similarly, $\hat{b}_2=1$, $\hat{b}_3=1$, and $\hat{b}_4=-1$. As seen in the example the Walsh code is used as spreading code. However, a variety of codes can be utilized for this purpose, which can be classified into two major categories.

- *PN codes*: pseudo-random noise code is a sequence of pulses which shows the appropriate features to be used in CDM. Although PN sequences look like noise, they can be exactly generated at both multiplexer and demultiplexer locally using a finite number of shift registers with a pre-defined initial state. The finite length of linear shift registers makes these codes deterministic. A local sequence has a high correlation with itself, but almost zero correlation with other sequences or a time-shifted version of itself. The term "pseudo" refers to the fact that a sequence starts to repeat a certain pattern after its period. In cryptography applications, to ensure security, using PN codes with very large period is a necessity. However, this is not a strict requirement for CDM.

- *Non-random orthogonal codes*: this kind of codes is designed in a specific and predefined manner and has a special set for desired length while satisfying primary features required by CDM. An instance of these codes is Walsh code shown in the previous example. Walsh code is used in the IS95/CDMA 2000 system.

As described, choosing a well-defined code plays a critical role in spreading process. The question may arise here is why the process of multiplying the code sequences to bits is called spreading process. To find the answer of this question, **Figure 8** illustrates an insightful example where the base-band information is converted to the transmitted signal via spreading code. Every bit in the information flow is represented by a pulse with T_b width in time domain. Hence, the effective bandwidth of the original data is proportional to $1/T_b$. Then, the information flow goes through spreading process where the pulses of data flow will be multiplied (in some systems it is XOR) by the spreading code sequence. Since the pulse duration for the spreading code is T_c which is smaller than the T_b, the generated signal by spreading signal is affected by T_c and it becomes a series of pulses with T_c width in time domain. The bandwidth of the transmitted signal is proportional to $1/T_c$ called chip rate. Therefore, the bandwidth of the transmitted signal is larger than original baseband signal. The ratio of spread spectrum and original base-band information is called spreading factor and can be expressed as follows.

$$Spreading\ Factor = \frac{chip\ rate}{bit\ rate} = \frac{T_b}{T_c} \tag{6}$$

Clearly, the spreading for the example illustrated in **Figure 8** is four since the chip rate is fourfold of bit rate.

Until this point, it seems very easy to decode the signal at the intended receiver by using the same spreading code with pre-known chip rate. The de-spreading process, i.e., multiplying the multiplexed signal with a spreading code needs a delicate requirement. At a node in the network, all the received signals multiplied by different spreading codes should be received with equal strength, otherwise the de-spreading process causes interference due to non-zero mutual correlation of spreading codes. This impedes recovering desired signal at that node. This problem is called near-far problem. The near far problem originally refers to a situation where reception of some strong signals makes it impossible to recover weak signals. Here, the unbalanced signal strength causes this intricate challenge. To cope with this problem, a power control mechanism is mandatory to ensure that all the signals from different sources will be received with equal strength. This matter brings two main challenges:

- High power consumption for far users: the power of a signal emitted on the air is attenuated by factor of d^{α} where d is the amount of distance the signal passes

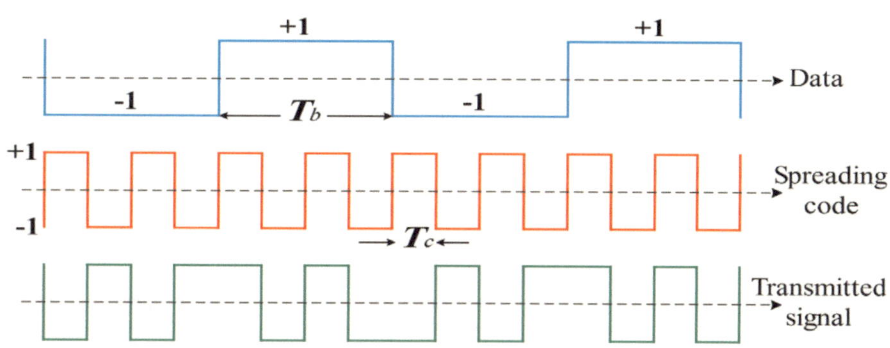

Figure 8.
Spreading process on $(-1,1,-1,1)$ via spreading code of $(-1,1,1,1)$ with spreading factor of 4.

and α is pathloss coefficient which depends on the surrounding area and object within there. For an open are and line of sight communication $\alpha \approx 2$. So, if a user wants to adjust its power to maintain a certain signal strength at destination, it takes the distance into the account. Hence, the emitted power of users far from the destination is pretty high. This is very probable in cellular network where some users may be located at cell-edge.

- Communication overhead and reduced overhead: to measure pathloss effect and adjust power, a sounding mechanism should be established to launch a two-way communication between source and destination. This sounding consumes available over-the-air time and reduces the overall throughput of the network.

Despite this challenge along with its other challenges like self-jamming and need for precise synchronization, CDM have shown some prominent advantages, as follows, paving the way for implementing it in several real-world communication.

- Efficient channelization and enhanced spectrum reuse
- Soft hand-off
- Immunity to interference
- Security

5. Power division multiplexing

Non-orthogonal multiplexing access (NOMA) is key enabler of the next generation wireless communication. Although the concept of NOMA is very broad, the power domain NOMA is the simplest and most popular NOMA. In contrast to orthogonal multiplexing access (OMA) approaches, like TDM and FDM, which separate the signals in frequency or time domain in order to avoid interference, NOMA embraces interference in both time and frequency domains. NOMA establishes multiple concurrent transmissions over the shared medium by adjusting the power levels of different signals.

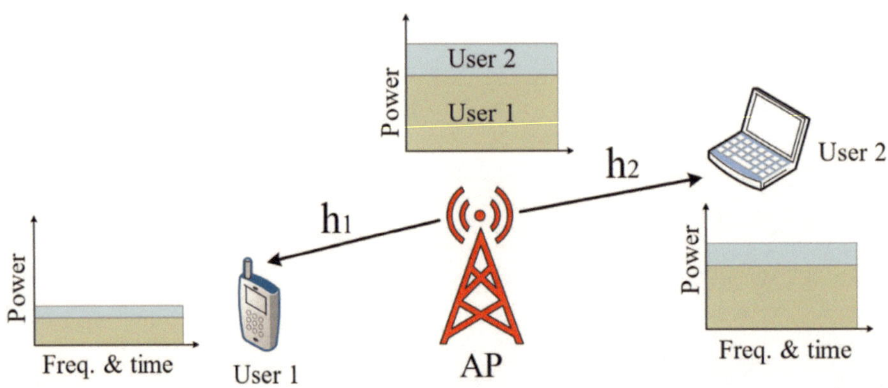

Figure 9.
An example of two-user network using power-domain NOMA in downlink transmission.

NOMA brings the massive connectivity, spectral efficiency, high throughput, and improved fairness all together and it is the key enabler of the fifth generation of wireless networks. By serving several users with available resources concurrently, it improves the connectivity and spectral efficiency. Also, it improves the network capacity, and prevent wasting the resources caused by assigning equal amount of resources to users with low data rate requirement or bad channel conditions. Another, brilliant feature of the NOMA is that it can be easily integrated into existing wireless communication technologies. For example, see its integration with LTE-A [12] and digital TV standard [13].

To illustrate how NOMA works, **Figure 9** shows a simple example of NOMA usage in a two-user network. In this example user 1 goes in the deep fade while user two hears the signal coming from the access point (AP) very clearly. All nodes are equipped with a single omni-directional antenna. Assume the channel between the AP and the ith user is h_i. Therefore, $h_1 \ll h_2$. The AP knows the global Channel State Information (CSI) perfectly. The AP intends to send message s_i to user i. To do so, it scales the s_i with power allocation factor α_i such that more power is assigned to the message of weak user. The superimposed message is as follows.

$$s_m = \sqrt{\alpha_1} s_1 + \sqrt{\alpha_2} s_2 \tag{7}$$

The received signal at the ith user is Y_i.

$$Y_i = s_m h_i + n_i = \sqrt{\alpha_1} s_1 h_i + \sqrt{\alpha_2} s_2 h_i + n_i \tag{8}$$

where n_i is additive white Gaussian noise. At the first user, i.e., weak user, the desired signal has a high power compared to the interference which is $\sqrt{\alpha_2} s_2 h_1$ to be decoded. This user decodes its desired signal by treating interference as noise. However, at the second user, i.e., strong user, the desired signal is drawn into strong interference. This user will pursue a decoding procedure called successive interference cancellation (SIC). Since the strong user knows the codebook used at the AP, it is able to decode interference, $\hat{s}_1$. Then, it subtracts the decoded interference from received signal. In the next step, the strong user endeavors to recover its desired signal from $Y_2^{(2)}$.

$$Y_2^{(2)} = \left(\sqrt{\alpha_1} s_1 h_2 - \sqrt{\alpha_1} \hat{s}_1 \hat{h}_2\right) + \sqrt{\alpha_2} s_2 h_2 + n_2 = e_1 + \sqrt{\alpha_2} s_2 h_2 + n_2 \tag{9}$$

If the channel is estimated perfectly and the interference is decoded flawless, error $e_1 = 0$, and the desired signal, s_2 can be recovered successfully. Assuming perfect CSI estimation, the sum-rate of the network is as follows.

$$R_{sum} = R_1 + R_2 = \log_2\left(1 + \frac{\alpha_1 |h_1|^2}{\alpha_2 |h_1|^2 + 1/\rho}\right) + \log_2\left(1 + \alpha_2 \rho |h_2|^2\right) \tag{10}$$

where $\rho = P_t/N_0$ and P_t is available power budget. This approach can be easily extended to $N-$user case. Assuming that channels' strength are sorted in ascending order, $|h_i|^2 < |h_j|^2$ for any $i < j$ and $i, j \in \{1, 2..., N\}$, the decoding order is $(1, 2, ..., N)$. It means the SIC at user l begins with decoding interference s_1, and then decodes the second powerful interference, i.e., s_2, and follows the interference subtraction until removing s_{i-1}. After subtracting all interferences which are stronger than desired signal, the user is capable of recovering its intended signal s_i. The R_{sum} can be expressed as follows.

$$R_{sum} = \sum_{i=1}^{N} \log_2 \left(1 + \frac{\alpha_i |h_i|^2}{\frac{1}{\rho} + \sum_{j=i+1}^{N} \alpha_j |h_i|^2} \right) \tag{11}$$

For better illustration of NOMA gain over conventional OMA schemes, like TDM and FDM, let us look at the following example.

Example 2: Consider two asymptotic cases with high power budget in two-user scenario: (i) the users' channel are equally strong $(h_1 = h_2 = h)$; (ii) one channel is very strong while the other one is very weak $(h_1 \ll h_2)$. The power is proportionally allocated to messages. The gain of power domain NOMA over OMA in each case is as follows.

Case 1: Let us say the power budget is P_T and $\rho = P_T/N_0$ where N_0 stands for variance of additive white Gaussian noise. In this case, the sum-capacity of network using OMA is C_{OMA}.

$$C_{OMA} = 0.5 \log_2\left(1 + \rho|h|^2\right) + 0.5 \log_2\left(1 + \rho|h|^2\right) \approx \log_2\left(\rho|h|^2\right) \tag{12}$$

where 0.5 coefficients lie in the fact that OMA shares the resources between users equally. In the NOMA approach, the whole spectrum is shared among both users. Due to power allocation strategy and equally strong channels $\alpha_1 = \alpha_2 = 1/2$. The sum-capacity of the network using the NOMA is C_{NOMA}.

$$\begin{aligned} C_{NOMA} &= \log_2\left(1 + \rho|h|^2\alpha_2\right) + \log_2\left(1 + |h|^2\alpha_1 \Big/ \left(|h|^2\alpha_2 + \frac{1}{\rho}\right)\right) \\ &\approx \log_2\left(1 + \rho|h|^2/2\right) + \log_2(2) \approx \log_2\left(\rho|h|^2\right) \end{aligned} \tag{13}$$

Therefore, OMA reaches to the performance of NOMA.

Case 2: In this case, $h_1 \ll h_2$ and $\rho|h_1|^2 \rightarrow 0$ and $\alpha_1 \rightarrow 1$.

$$C_{OMA} = 0.5 \log_2\left(1 + \rho|h_1|^2\right) + 0.5 \log_2\left(1 + \rho|h_2|^2\right) \approx 0.5 \log_2\left(1 + \rho|h_2|^2\right) \tag{14}$$

$$\begin{aligned} C_{NOMA} &= \log_2\left(1 + \rho|h_2|^2\alpha_2\right) + \log_2\left(1 + |h_1|^2\alpha_1 \Big/ \left(|h_2|^2\alpha_2 + \frac{1}{\rho}\right)\right) \\ &\approx \log_2\left(1 + \rho|h_2|^2\alpha_2\right) + \log_2\left(1 + \frac{\alpha_1}{\alpha_2}\right) = \log_2\left(\rho|h_2|^{2.}\right) \end{aligned} \tag{15}$$

Obviously, in the latter case, $C_{NOMA} = 2 \times C_{OMA}$. Based on these bounds, the performance of the NOMA for two users can be expressed as follows:

$$C_{OMA} \le C_{NOMA} \le 2 \times C_{OMA} \tag{16}$$

The previous example shows that the performance of NOMA is highly dependent on the channel difference among serving users. Therefore, when an AP serves many active users, it is very critical to employ a grouping mechanism to pair a strong and weak user among all possible choices such that the overall throughput of the network is maximized. Each pair will be served separately in different time slot. When the AP has just one antenna, best strategy is measuring the magnitude of the channels of different users, sorting the users based on their channel strength, and then choosing a N-user group in which the channel strength difference among two adjacent users in decoding order is larger than a certain threshold.

To put it briefly, the performance of NOMA depends on its three key components: (i) grouping mechanism; (ii) power allocation scheme; and (iii) SIC at all user except the weakest one. While the grouping mechanism aims at finding best pairs among all users to be fed into power allocator, the power allocation scheme should assign a portion of available power to intended messages of user included in the pairs such that the minimum required data rate of all users is met and the decoding order at the user side is preserved. At the AP, depending on the application, there are two main strategies for power allocation. The first one is maximization of throughput using all available power budget. The second strategy is minimization of the power consumption while requirements of all users are met. After a proper power allocation, all users except the weakest one should be able to recover all strong interference based on decoding order through and accurate and reliable SIC. Although NOMA shows a promising performance theoretically, it faces several challenges in practice. These challenges are described in what follows.

- *Privacy of weak users:* the first and foremost problem of the NOMA is privacy of weak user since their message will be decoded by all users possessing higher decoding order. Indeed, at the end of a cycle of NOMA, the ith user knows the messages of all weaker user, i.e., s_j if $|h_j| < |h_i|$. For example, after SIC completion, the strongest user identifies all the messages sent by the AP for other users

- *Long processing delay for strong users*: in SIC approach, a user first removes all the interference messages stronger than desired signal. This inflicts a huge amount of computational complexity and long processing delay to strong users which must iterate this process several times.

- *Error propagation*: the performance of SIC is highly intertwined with accurate interference reconstruction on subtraction. This fact mandates reliable channel estimation and accurate interference decoding. Failure in these steps for one of strong interferences will introduce an error which impacts the SIC operation for decoding and removing the weaker interferences at the same user. Accumulation of errors within the whole process makes it very difficult to recover the desired signal polluted by errors and residual interferences.

- *Non-trivial grouping*: although the grouping mechanism seems very straightforward for single-antenna AP and single-antenna users (SISO network), the channel strength concept becomes ambiguous for more complicated scenarios, like MISO and MIMO. For example, in MISO case, the channel between a user, say user i, and one of AP's antennas may be so weak while the channel of the same user and other antennas at the AP is very strong. In these cases, to sort the users' strength, the different measures are taken into account, like distance from AP, norm of the channel vector, etc. These metrics stem from theoretical aspects of wireless channels and may show unacceptable outcome when implemented in practice.

- *Sub-optimal performance:* at the AP side, power allocation problem becomes a non-trivial task for large pairs and nodes with multiple antenna (see, e.g., [14]). Due to several requirements imposed by either user or nature of SIC, the problem is a non-convex problem with several variable. In almost all cases, there is no closed-form solution for optimal power allocation. To reach a solution, the problem will be relaxed into a simpler problem, or a heuristic algorithm reaches to a near optimal solution. So, the power allocation problem entails high computational cost and may yields a non-optimal solution which diminish the gain of the NOMA over OMA.

Although, the term NOMA has referred to power domain multiplexing through this section, the concept of NOMA is much broader and includes other kind of schemes which are totally different from presented one. Actually, in this section, the power domain NOMA is presented. To read more about other variation of NOMA, see [15, 16].

6. Conclusion

In this chapter, we briefly studied the important multiplexing/de-multiplexing techniques, including TDM, CDM, FDM, OFDM, and power-domain NOMA. These techniques play important roles in the past, current, and future generations of wireless communication networks. The main objectives of these techniques are providing the opportunity of concurrent multi-user transmissions and effective utilization of available communication resources. However, the multiplexing approaches are not limited to the investigated techniques, and other techniques, such as spatial multiplexing, delay-Doppler spread multiplexing, pattern division multiplexing can be found in the communication territory. Also, it is very common in communication technologies to employ more than one multiplexing technique for transmission.

Author details

Pedram Kheirkhah Sangdeh and Huacheng Zeng*
University of Louisville, Louisville, KY, USA

*Address all correspondence to: huacheng.zeng@louisville.edu

References

[1] IEEE Standard for Local and metropolitan area networks Part 16. Air Interface for Broadband Wireless Access Systems Amendment 3: Advanced Air Interface. IEEE Std 802.16m-2011 (Amendment to IEEE Std 802.16-2009). 2011. pp. 1-1112. DOI: 10.1109/IEEESTD.2011.5765736

[2] Li Y. Tactical data link and its application technology. Journal of China Academy of Electronics and Information Technology. 2007;**2**(2): 211-217

[3] Bluetooth SI. Bluetooth core specification version 4.0. Specification of the Bluetooth System. July 2010

[4] ETSI: Broadband Radio Access Networks (BRAN); HIPERLAN type 2 technical specification; physical (PHY) layer. August. 1999

[5] ATM Forum, editor. ATM User-network Interface (UNI) Specification Version 3.1. Foster City, CA, USA: Prentice Hall; 1995

[6] IEEE 802.11n 2009. Part 11: Wireless LAN medium access control (MAC) and physical layer (PHY) specifications, amendment 5: Enhancement for higher throughput. October 2009

[7] IEEE 802.11p Draft Standard for Information Technology, IEEE Standard. Vol. 802.11. 2009

[8] IEEE 802.11ac—Amendment 4: Enhancements for Very High Throughput for Operation in Bands below 6 GHz. IEEE P802.11ac/D5.0. 2013

[9] Sesia S, Baker M, Toufik I. LTE-The UMTS Long Term Evolution: From Theory to Practice. New York, NY, USA: John Wiley & Sons; 2011

[10] Dahlman E, Parkvall S, Skold J. 4G: LTE/LTE-Advanced for Mobile Broadband. New York, NY, USA: Academic Press; 2013

[11] Ghosh A, Ratasuk R, Mondal B, Mangalvedhe N, Thomas T. LTE-advanced: Next-generation wireless broadband technology. IEEE Wireless Communications. 2010;**17**(3):10-22

[12] 3rd Generation Partnership Project (3GPP). Study on downlink multiuser superposition transmission for LTE. 2015

[13] Zhang L, Li W, Wu Y, Wang X, Park SI, Kim HM, et al. Layered-division multiplexing: Theory and practice. IEEE Transactions on Broadcasting. 2016; **62**(1):216-232. DOI: 10.1109/TBC.2015.2505408

[14] Wang J, Peng Q, Huang Y, Wang H, You X. Convexity of weighted sum rate maximization in NOMA systems. IEEE Signal Processing Letters. 2017;**24**(9): 1323-1327. DOI: 10.1109/LSP.2017.2722546

[15] Ding Z, Lei X, Karagiannidis GK, Schober R, Yuan J, Bhargava VK. A survey on non-orthogonal multiple access for 5G networks: Research challenges and future trends. IEEE Journal on Selected Areas in Communications. 2017;**35**(10): 2181-2195. DOI: 10.1109/JSAC.2017.2725519

[16] Chen S, Sun S, Kang S, Ren B. Pattern division multiple access (PDMA). In: Multiple Access Techniques for 5G Wireless Networks and Beyond. Cham: Springer; 2019. pp. 451-492

Section 3

MIMO and 5G

Chapter 4

Coherent Receiver for Turbo Coded Single-User Massive MIMO-OFDM with Retransmissions

K. Vasudevan, Shivani Singh and A. Phani Kumar Reddy

Abstract

Single-user massive multiple-input multiple-output (MIMO) systems have a large number of antennas at the transmitter and receiver. This results in a large overall throughput (bit-rate), of the order of tens of gigabits per second, which is the main objective of the recent fifth-generation (5G) wireless standard. It is feasible to have a large number of antennas in mm-wave frequencies, due to the small size of the antennas. This chapter deals with the coherent detection of orthogonal frequency division multiplexed (OFDM) signals transmitted through frequency-selective Rayleigh fading MIMO wireless channels. Low complexity, discrete-time algorithms are developed for channel estimation, carrier and timing synchronization, and finally turbo decoding of the data at the receiver. Computer simulation results are presented to validate the theory.

Keywords: 5G, channel capacity, channel estimation, single-user massive MIMO, OFDM, spatial multiplexing, retransmissions, synchronization, turbo codes

1. Introduction

The main objective of the fifth-generation [1–15] wireless communication standard is to provide peak data rates of 10 gigabit per second (Gbps) for each user, ultra-low latency (the time duration between transmission of information and getting a response) of less than 1 ms, and, last but not the least, very low bit error rates (BER) ($<10^{-10}$). High data rates are essential for streaming ultrahigh definition (4k) video. Low latency is required for future driverless cars and remote surgeries. An important feature of the 5G network is that it involves not only people but also smart devices. For example, it may be possible to control a microwave oven or geyser located in the home, from the office. High data rates are feasible by using a large number of transmitting antennas. For example, if each transmit antenna transmits at a rate of 100 megabits per second (Mbps), then using 100 transmit antennas would result in an overall bit-rate of 10 Gbps. This technique of increasing the overall bit-rate by using a large number of transmit antennas is also known as spatial multiplexing (not to be confused with spatial modulation [16–20], wherein not all the transmit antennas are simultaneously active). This is illustrated in **Figure 1**, where the ith transmit antenna sends C_i bits of information and each of the receive antennas gets C/N bits

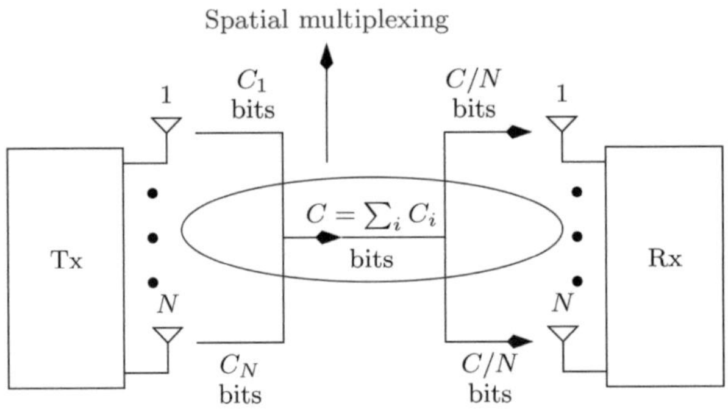

Figure 1.
Illustration of spatial multiplexing for $N \times N$ MIMO.

of information, in each transmission (see *Proposition A.1* and *A.2* in [21]). It must be noted that a large array of transmit antennas can also be used for beamforming [22, 23] and beam steering (the ability to focus the transmitted signal in a particular direction, without moving the antenna), which is not the topic of this chapter. In fact, the basic idea used in this chapter is captured in the following proposition.

Proposition 1.1 *Signals transmitted and received by antennas separated by at least $\lambda/2$ ($\lambda = c/\nu$ where c is the velocity of light and ν is the carrier frequency) undergo independent fading.*

A typical massive MIMO antenna array is shown in **Figure 2**. The black dots denote the antennas, and the circles denote obstructions used to prevent mutual coupling between the antennas. While spatial multiplexing is a big advantage in massive MIMO, the main problem lies in the high complexity of data detection at the receiver. To understand this issue, consider the signal model:

$$\tilde{\mathbf{R}} = \tilde{\mathbf{H}}\mathbf{S} + \tilde{\mathbf{W}} \tag{1}$$

where $\tilde{\mathbf{R}} \in \mathbb{C}^{N\times 1}$ is the received vector, $\tilde{\mathbf{H}} \in \mathbb{C}^{N\times N}$ is the channel matrix, $\mathbf{S} \in \mathbb{C}^{N\times 1}$ is the symbol vector drawn from an M-ary 2D constellation, and $\tilde{\mathbf{W}} \in \mathbb{C}^{N\times 1}$ is the

Figure 2.
A massive MIMO antenna array.

additive white Gaussian noise (AWGN) vector. Here $\mathbb{C}$ denotes the set of complex numbers. Due to Proposition 1.1, the elements of $\tilde{\mathbf{H}}$ are statistically independent. Moreover, if there is no line-of-sight (LOS) path between the transmitter and receiver, the elements of $\tilde{\mathbf{H}}$ are zero-mean Gaussian. The elements of $\tilde{\mathbf{W}}$ are also assumed to be independent. The real and imaginary parts of the elements of $\tilde{\mathbf{H}}$ and $\tilde{\mathbf{W}}$ are also assumed to be independent. Now, the problem statement is find $\mathbf{S}$ given $\tilde{\mathbf{R}}$. There are several methods of solving this problem, assuming that $\tilde{\mathbf{H}}$ is known.

1. Perform an exhaustive search over all the M^N possibilities of $\mathbf{S}$. This is known as the maximum likelihood (ML) approach, which has got an exponential complexity.

2. Pre-multiply $\tilde{\mathbf{R}}$ with $\tilde{\mathbf{H}}^{-1}$. This is known as the zero-forcing approach and has a complexity of the order of $2N^3$ (N^3 complexity for computing the inverse and another N^3 for matrix multiplication). This approach usually leads to noise enhancement and a poor symbol error rate (SER) performance.

3. The third approach, known as sphere decoding, has polynomial complexity ($\sim C_0 \times N^{C_1}$, where C_0>1 and C_1>3) and has been widely studied in the literature [24–33].

Data detection in single-user massive MIMO systems using retransmissions, having a complexity of $N_{rt} \times N^3$, where N_{rt} is the number of retransmissions, has been proposed [34], where it was assumed that $\tilde{\mathbf{H}}$ is known at the receiver. In this work, which is an extension of [34], we present a coherent receiver for massive MIMO systems, where not only $\tilde{\mathbf{H}}$ but also the carrier frequency offset and timing are estimated. Moreover, the signal model in Eq. (1) is valid for flat fading channels. When the channel is frequency selective (the length of the discrete-time channel impulse response is greater than unity), orthogonal frequency division multiplexing needs to be used, since OFDM converts a frequency-selective channel into a flat fading channel (length of the discrete-time channel impulse response is equal to unity) [35]. To this end, the channel estimation and carrier and timing synchronization algorithms developed in [36] for single-input single-output (SISO) OFDM, [37, 38] for single-input multiple-output (SIMO) OFDM, and [21, 39] for multiple-input multiple-output (MIMO) OFDM are used in this work. In [40], a linear prediction-based detection of serially concatenated QPSK is presented, which does not require any preamble. The prospect of using superimposed training [41] in the context of massive MIMO looks quite intimidating, since the signal at each receive antenna is already a superposition of the signals from a large number of transmit antennas.

This work is organized as follows. Section 2 presents the system model. The discrete-time receiver algorithms are presented in Section 3. The computer simulation results are discussed in Section 4, and the chapter concludes with Section 5.

2. System model

The transmitted frame structure is shown in **Figure 3(a)**. The signal in the blue boxes is sent from transmit antenna n_t. The signal in the red boxes is sent from other antennas. Note that in the preamble phase, only one transmit antenna is active at a time, whereas in the data phase, all transmit antennas are active simultaneously. In

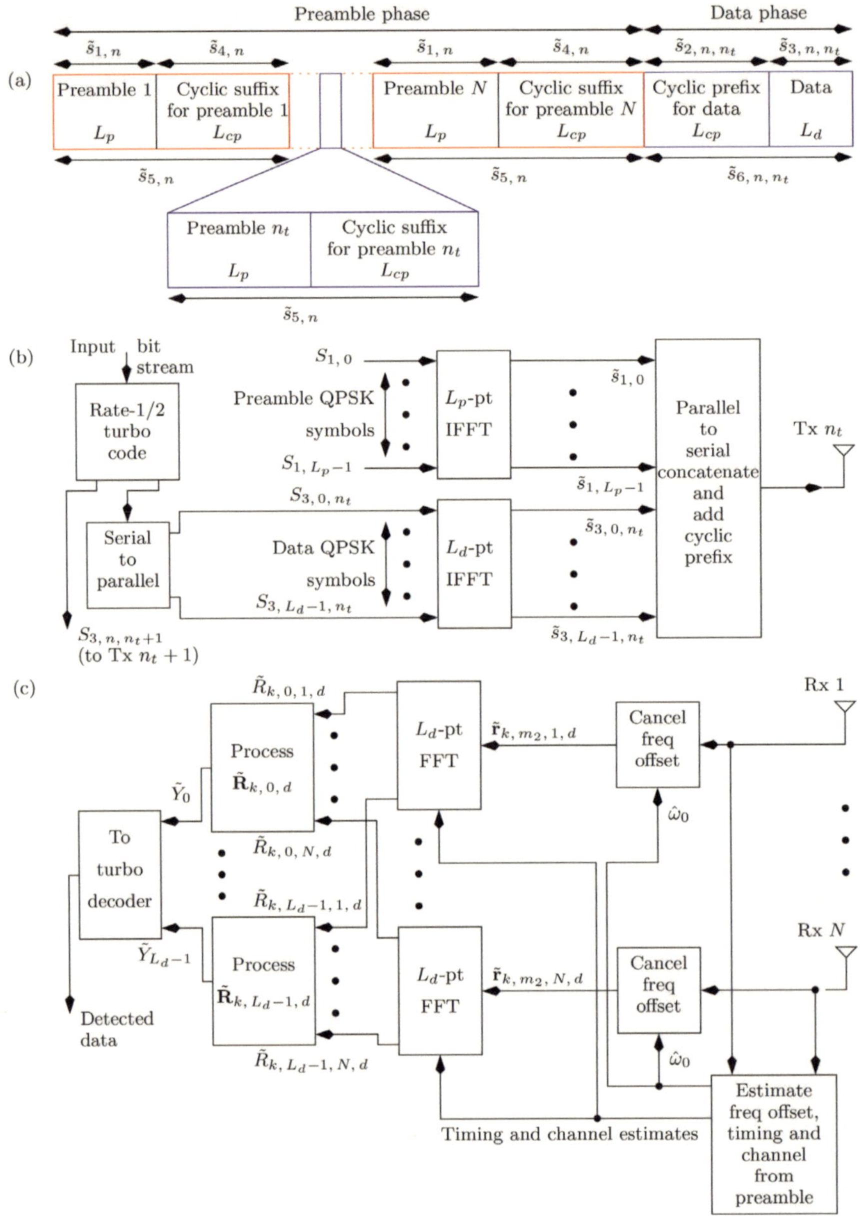

Figure 3.
(a) Frame structure for k^{th} retransmission. (b) Signal from transmit antenna n_t. (c) Receiver for the data phase.

practice, each transmit antenna could use a different preamble. However, in this work, we assume that all transmit antennas use the same preamble. The signals in **Figure 3(a)** are defined as follows (similar to [21]):

$$
\begin{aligned}
\tilde{s}_{1,n} &= \frac{1}{L_p}\sum_{i=0}^{L_p-1} S_{1,i} e^{j2\pi ni/L_p} \quad \text{for } 0 \le n \le L_p - 1 \\
\tilde{s}_{3,n,n_t} &= \frac{1}{L_d}\sum_{i=0}^{L_d-1} S_{3,i,n_t} e^{j2\pi ni/L_d} \quad \text{for } 0 \le n \le L_d - 1 \\
\tilde{s}_{2,n,n_t} &= \tilde{s}_{3,L_d-L_{cp}+n,n_t} \quad \text{for } 0 \le n \le L_{cp} - 1 \\
\tilde{s}_{4,n} &= \tilde{s}_{1,n} \quad \textit{for } 0 \le n \le L_{cp} - 1.
\end{aligned} \tag{2}
$$

The term i in the above equations denotes the ith subcarrier, n denotes the time index, and $1 \le n_t \le N$ is the index to the transmit antenna. Note that in this work, the same preamble is transmitted one after the other by each of the transmit antennas, as shown in **Figure 3(a)**. In [21], different preambles are transmitted simultaneously from all the transmit antennas. The channel coefficients $\tilde{h}_{k,n,n_r,n_t}$ associated with the receive antenna n_r $(1 \le n_r \le N)$ and transmit antenna n_t $(1 \le n_t \le N)$ for the kth retransmission are $\mathscr{CN}\left(0, 2\sigma_f^2\right)$ ($\mathscr{CN}(\cdot)$ denotes a circularly symmetric Gaussian random variable) and satisfy the following relations [21]:

$$\begin{aligned} \frac{1}{2}E\left[\tilde{h}_{k,n,n_r,n_t}\tilde{h}^*_{k,m,n_r,n_t}\right] &= \sigma_f^2\delta_K(n-m) \\ \frac{1}{2}E\left[\tilde{h}_{k,n,n_r,n_t}\tilde{h}^*_{k,n,m_r,n_t}\right] &= \sigma_f^2\delta_K(n_r-m_r) \\ \frac{1}{2}E\left[\tilde{h}_{k,n,n_r,n_t}\tilde{h}^*_{k,n,n_r,m_t}\right] &= \sigma_f^2\delta_K(n_t-m_t) \\ \frac{1}{2}E\left[\tilde{h}_{k,n,n_r,n_t}\tilde{h}^*_{i,n,n_r,n_t}\right] &= \sigma_f^2\delta_K(k-i) \end{aligned} \tag{3}$$

where "*" denotes complex conjugate and $\delta_K(\cdot)$ is the Kronecker delta function. Observe that Eq. (3) implies a uniform power delay profile. Even though an exponential power delay profile is more realistic, we have used a uniform power delay profile, since it is expected to give the worst-case BER performance, as all the multipath components have the same power [21]. The channel is assumed to be quasi-static, that is, $\tilde{h}_{k,n,n_r,n_t}$ is time-invariant over one frame (retransmission). The length of all the N^2 channel impulse responses is assumed to be L_h, which is proportional to the difference between the longest and shortest multipath [21]. The channel span assumed by the receiver is [21, 36, 39].

$$L_{hr} = 2L_h - 1. \tag{4}$$

The length of the cyclic prefix or suffix is [21, 36, 39].

$$L_{cp} = L_{hr} - 1. \tag{5}$$

The length of the preamble is L_p, and the length of the data is L_d. The AWGN noise samples $\tilde{w}_{k,n,n_r}$ for the kth retransmission at time n and receive antenna n_r are $\mathscr{CN}\left(0, 2\sigma_w^2\right)$ and satisfy

$$\begin{aligned} \frac{1}{2}E\left[\tilde{w}_{k,n,n_r}\tilde{w}^*_{k,m,n_r}\right] &= \sigma_w^2\delta_K(n-m) \\ \frac{1}{2}E\left[\tilde{w}_{k,n,n_r}\tilde{w}^*_{k,n,m_r}\right] &= \sigma_w^2\delta_K(n_r-m_r) \\ \frac{1}{2}E\left[\tilde{w}_{k,n,n_r}\tilde{w}^*_{i,m,n_r}\right] &= \sigma_w^2\delta_K(k-i). \end{aligned} \tag{6}$$

The noise and channel coefficients are assumed to be independent. The frequency offset ω_0 is uniformly distributed over $[-0.03, 0.03]$ radians, and the ML frequency offset estimator searches in the range $[-\omega_{0,\max}, \omega_{0,\max}]$ radians [42] where

$$\omega_{0,\max} = 0.04\ radian. \tag{7}$$

For convenience, and without loss of generality, we assume that ω_0 is constant over N_{rt} retransmissions.

During the preamble phase, the signal at receive antenna n_r, for the kth retransmission, can be written as (for $0 \le n \le L_p + L_{cp} + L_h - 2$)

$$\begin{aligned}\tilde{r}_{k,n,n_r,n_t,p} &= \left(\tilde{s}_{5,n} \star \tilde{h}_{k,n,n_r,n_t}\right) \mathrm{e}^{\mathrm{j}\omega_0 n} + \tilde{w}_{k,n,n_r,n_t,p} \\ &= \tilde{y}_{k,n,n_r,n_t,p} \mathrm{e}^{\mathrm{j}\omega_0 n} + \tilde{w}_{k,n,n_r,n_t,p}\end{aligned} \tag{8}$$

where " $\star$ " denotes linear convolution, $\tilde{s}_{5,n}$ is depicted in **Figure 3(a)**, $\tilde{h}_{k,n,n_r,n_t}$ denotes the channel impulse response between transmit antenna n_t and receive antenna n_r for the kth retransmission, and

$$\tilde{y}_{k,n,n_r,n_t,p} = \tilde{s}_{5,n} \star \tilde{h}_{k,n,n_r,n_t}. \tag{9}$$

The subscript "p" in Eqs. (8) and (9) denotes the preamble. Note that any random carrier phase can be absorbed in the channel impulse response. We have

$$\begin{aligned}\tilde{s}_{1,n} \odot_{L_p} \tilde{s}^*_{1,-n} &= E_s \delta_K(n) \\ &\overset{L_p}{\rightleftharpoons} \left|\tilde{S}_{1,i}\right|^2 \\ &= \text{a constant for } 0 \le i \le L_p - 1\end{aligned} \tag{10}$$

where " $\odot_{L_p}$ " denotes an L_p-point circular convolution, "$\overset{L_p}{\rightleftharpoons}$" denotes the L_p-point discrete Fourier transform (DFT) or the fast Fourier transform (FFT), and

$$E_s = \sum_{n=0}^{L_p-1} \left|\tilde{s}_{1,n}\right|^2. \tag{11}$$

Due to the presence of the cyclic suffix, we have

$$\tilde{s}_{5,n} \star \tilde{s}^*_{1,-n} \begin{cases} = 0 & \text{for } 1 \le n \le L_{hr} - 1 \\ = E_s & \text{for } n = 0 \\ \ll E_s & \text{otherwise.} \end{cases} \tag{12}$$

Assuming perfect carrier and timing synchronization (ω_0 is perfectly canceled and the frame boundaries are perfectly known) at the receiver, the signal at the output of the L_p-point FFT for the ith $(0 \le i \le L_p - 1)$ subcarrier and receive antenna n_r, due to the preamble sent from transmit antenna n_t during the kth retransmission, is

$$\tilde{R}_{k,i,n_r,n_t,p} = \tilde{H}_{k,i,n_r,n_t} S_{1,i} + \tilde{W}_{k,i,n_r,n_t,p} \tag{13}$$

where $\tilde{H}_{k,i,n_r,n_t}$ and $\tilde{W}_{k,i,n_r,n_t,p}$ denote the L_p-point FFT of $\tilde{h}_{k,n,n_r,n_t}$ and $\tilde{w}_{k,n,n_r,n_t,p}$, respectively. The average SNR per bit corresponding to Eq. (13) is

$$\begin{aligned} SNR_{\mathrm{av},\,b,\,p} &= \frac{E\left[2A^2\left|\tilde{H}_{k,\,i,\,n_r,\,n_t}\right|^2\right]}{E\left[\left|\tilde{W}_{k,\,i,\,n_r,\,n_t,\,p}\right|^2\right]} \times \frac{NN_{rt}}{2} \\ &= \frac{2A^2\left(2L_h\sigma_f^2\right)NN_{rt}}{\left(2L_p\sigma_w^2\right) \times 2} \\ &= \frac{A^2L_h\sigma_f^2NN_{rt}}{L_p\sigma_w^2} \end{aligned} \tag{14}$$

where [36]

$$\begin{aligned} E\left[\left|\tilde{H}_{k,\,i,\,n_r,\,n_t}\right|^2\right] &= 2L_h\sigma_f^2 \\ E\left[\left|\tilde{W}_{k,\,i,\,n_r,\,n_t,\,p}\right|^2\right] &= 2L_p\sigma_w^2 \\ E\left[\left|S_{1,\,i}\right|^2\right] &\triangleq 2A^2 \\ &= E_s \end{aligned} \tag{15}$$

where A is a constant to be determined and it is assumed that each sample of each receive antenna gets $2/(NN_{rt})$ bits of information during the preamble phase (see *Proposition A.2* in [21]).

During the data phase, the signal for the kth retransmission at receive antenna n_r can be written as (for $0 \le n \le L_d + L_{cp} + L_h - 2$)

$$\begin{aligned} \tilde{r}_{k,\,n,\,n_r,\,d} &= \sum_{n_t=1}^{N}\left(\tilde{s}_{6,\,n,\,n_t} \star \tilde{h}_{k,\,n,\,n_r,\,n_t}\right)\mathrm{e}^{\mathrm{j}\omega_0 n} + \tilde{w}_{k,\,n,\,n_r,\,d} \\ &= \tilde{y}_{k,\,n,\,n_r,\,d}\mathrm{e}^{\mathrm{j}\omega_0 n} + \tilde{w}_{k,\,n,\,n_r,\,d} \end{aligned} \tag{16}$$

where $\tilde{s}_{6,\,n,\,n_t}$ is depicted in **Figure 3(a)** and

$$\tilde{y}_{k,\,n,\,n_r,\,d} = \sum_{n_t=1}^{N} \tilde{s}_{6,\,n,\,n_t} \star \tilde{h}_{k,\,n,\,n_r,\,n_t}. \tag{17}$$

The subscript "d" in Eqs. (16) and (17) denotes data. Assuming perfect carrier and timing synchronization at the receiver, the signal at the output of the L_d-point FFT for the ith $(0 \le i \le L_d - 1)$ subcarrier and receive antenna n_r, during the kth retransmission, is

$$\tilde{R}_{k,\,i,\,n_r,\,d} = \sum_{n_t=1}^{N} \tilde{H}_{k,\,i,\,n_r,\,n_t} S_{3,\,i,\,n_t} + \tilde{W}_{k,\,i,\,n_r,\,d} \tag{18}$$

where $\tilde{H}_{k,\,i,\,n_r,\,n_t}$ and $\tilde{W}_{k,\,i,\,n_r,\,d}$ denote the L_d-point FFT of $\tilde{h}_{k,\,n,\,n_r,\,n_t}$ and $\tilde{w}_{k,\,n,\,n_r,\,d}$, respectively. The average SNR per bit corresponding to Eq. (18) is

$$\begin{aligned} SNR_{\text{av}, b, d} &= \frac{E\left[\left|\sum_{n_t=1}^{N} \tilde{H}_{k, i, n_r, n_t} S_{3, i, n_t}\right|^2\right]}{E\left[\left|\tilde{W}_{k, i, n_r, d}\right|^2\right]} \times 2N_{rt} \\ &= \frac{2\left(2L_h\sigma_f^2\right)N(2N_{rt})}{2L_d\sigma_w^2} \\ &= \frac{4L_h\sigma_f^2 NN_{rt}}{L_d\sigma_w^2} \end{aligned} \tag{19}$$

where [36].

$$\begin{aligned} E\left[\left|\tilde{W}_{k, i, n_r, d}\right|^2\right] &= 2L_d\sigma_w^2 \\ E\left[\left|S_{3, i, n_t}\right|^2\right] &\triangleq 2 \end{aligned} \tag{20}$$

and it is assumed that each receive antenna gets $1/(2N_{rt})$ bits of information in each transmission [34]. We impose the constraint that

$$\begin{aligned} &\text{SNR}_{\text{av}, b, p} = \text{SNR}_{\text{av}, b, d} \\ &\Rightarrow \frac{A^2 L_h\sigma_f^2 NN_{rt}}{L_p\sigma_w^2} = \frac{4L_h\sigma_f^2 NN_{rt}}{L_d\sigma_w^2} \\ &\Rightarrow A = \sqrt{\frac{4L_p}{L_d}}. \end{aligned} \tag{21}$$

Let us now compare the average power of the preamble with that of the data, at the transmitter. The average power of the preamble in the time domain is

$$\begin{aligned} E\left[\left|\tilde{s}_{1, n}\right|^2\right] &= \frac{1}{L_p^2} E\left[\sum_{i=0}^{L_p-1} S_{1, i} e^{j2\pi ni/L_p} \sum_{l=0}^{L_p-1} S_{1, l}^* e^{-j2\pi nl/L_p}\right] \\ &= \frac{1}{L_p^2} \sum_{i=0}^{L_p-1} \sum_{l=0}^{L_p-1} E\left[S_{1, i} S_{1, l}^*\right] \times e^{-j2\pi n(i-l)/L_p} \\ &= \frac{1}{L_p^2} \sum_{i=0}^{L_p-1} \sum_{l=0}^{L_p-1} 2A^2 \delta_K(i-l) \times e^{-j2\pi n(i-l)/L_p} \\ &= \frac{2A^2}{L_p} \\ &= \frac{8}{L_d} \end{aligned} \tag{22}$$

where A is defined in Eqs. (15) and (21). Similarly, the average power of the data in the time domain is

$$E\left[\left|\tilde{s}_{3, n, n_t}\right|^2\right] = \frac{2}{L_d}. \tag{23}$$

Therefore, the radio frequency (RF) amplifiers at the transmitter must have a dynamic range of at least (note that the RF amplifiers have to also deal with the peak-to-average power ratio (PAPR) problem [43–49])

$$\begin{aligned} 10\log_{10}\left(\frac{E\left[\left|\tilde{s}_{1,n}\right|^2\right]}{E\left[\left|\tilde{s}_{3,n,n_t}\right|^2\right]}\right) &= 10\log_{10}(4)\ \text{dB} \\ &= 6\ \text{dB}. \end{aligned} \tag{24}$$

Let us now consider the case where the preamble power is equal to the data power at each transmit antenna. From Eqs. (22) and (23), we have [36]

$$\begin{aligned} \frac{2A^2}{L_p} &= \frac{2}{L_d} \\ \Rightarrow A &= \sqrt{\frac{L_p}{L_d}}. \end{aligned} \tag{25}$$

Substituting for A from Eq. (25), we obtain the average SNR per bit of the preamble phase and the data phase as

$$\begin{aligned} \text{SNR}_{\text{av},b,p} &= \frac{A^2 L_h \sigma_f^2 N N_{rt}}{L_p \sigma_w^2} \\ &= \frac{L_h \sigma_f^2 N N_{rt}}{L_d \sigma_w^2} \\ \text{SNR}_{\text{av},b,d} &= \frac{4 L_h \sigma_f^2 N N_{rt}}{L_d \sigma_w^2} \\ \Rightarrow 10\log_{10}\left(\frac{\text{SNR}_{\text{av},b,d}}{\text{SNR}_{\text{av},b,p}}\right) &= 10\log_{10}(4)\ \text{dB}. \\ &= 6\ \text{dB}. \end{aligned} \tag{26}$$

In other words, the average SNR per bit of the preamble phase would be less than that of the data phase by 6 dB. In what follows, we assume that A is given by (21).

3. Receiver algorithms

The receiver algorithms have been adapted from [21, 36, 37, 39] and will be briefly described in the following subsections.

3.1 Start of frame and frequency offset estimation

The first task of the receiver is to detect the presence of a valid signal, that is, the start of frame (SoF). The SoF detection and coarse frequency offset estimation are performed for each receive antenna $1 \le n_r \le N$, transmit antenna $1 \le n_t \le N$, and retransmission $1 \le k \le N_{rt}$ as given by the following rule (similar to Eq. (17) in [21]: choose that value of $\hat{m}_k(n_r, n_t)$ and $\hat{\nu}_k(n_r, n_t)$ which maximizes

$$\left|\left(\tilde{r}_{k,\,m,\,n_r,\,n_t,\,p}\,\mathrm{e}^{-\mathrm{j}\hat{\nu}_k(n_r,\,n_t)m}\right) \star \tilde{s}^{*}_{1,\,L_p-1-m,\,n_t}\right| \tag{27}$$

where $\tilde{r}_{k,\,m,\,n_r,\,n_t,\,p}$ is given in Eq. (8) and

$$\hat{\nu}_k(n_r, n_t) \in \left\{-\omega_{0,\,\max} + \frac{2l\omega_{0,\,\max}}{B_1}\right\} \tag{28}$$

for $0 \le l \le B_1$, where l and B_1 [21] are positive integers and $\omega_{0,\,\max}$ is given in Eq. (7). Observe that $\hat{m}_k(n_r, n_t)$ satisfies Eqs. (18) and (19) in [21]. The average value of the frequency offset estimate is given by

$$\hat{\omega}_0 = \frac{\sum_{k=1}^{N_{rt}}\sum_{n_r=}^{N}\sum_{n_t=1}^{N}\hat{\nu}_k(n_r, n_t)}{N^2 N_{rt}}. \tag{29}$$

3.2 Channel estimation

We assume that the SoF has been estimated using Eq. (27) with outcome m_0 given by (assuming the condition (19) in [21] is satisfied for all k, n_r, and n_t)

$$m_0 = \hat{m}_1(1, 1) - L_p + 1 \quad 0 \le m_0 \le L_h - 1 \tag{30}$$

and the frequency offset has been perfectly canceled [36, 38]. Observe that any value of k, n_r, and n_t can be used in the computation of Eq. (30). We have taken $k = n_r = n_t = 1$. Define [21, 36, 39].

$$m_1 = m_0 + L_h - 1. \tag{31}$$

The steady-state, preamble part of the received signal for the kth retransmission and receive antenna n_r can be written as [21, 36, 39]

$$\tilde{\mathbf{r}}_{k,\,m_1,\,n_r,\,n_t,\,p} = \tilde{\mathbf{s}}_5\tilde{\mathbf{h}}_{k,\,n_r,\,n_t} + \tilde{\mathbf{w}}_{k,\,m_1,\,n_r,\,n_t,\,p} \tag{32}$$

where

$$\begin{aligned}
\tilde{\mathbf{r}}_{k,\,m_1,\,n_r,\,n_t,\,p} &= \left[\tilde{r}_{k,\,m_1,\,n_r,\,n_t,\,p} \;\cdots\; \tilde{r}_{k,\,m_1+L_p-1,\,n_r,\,n_t,\,p}\right]^T_{L_p\times 1}\\
\tilde{\mathbf{w}}_{k,\,m_1,\,n_r,\,n_t,\,p} &= \left[\tilde{w}_{k,\,m_1,\,n_r,\,n_t,\,p} \;\cdots\; \tilde{w}_{k,\,m_1+L_p-1,\,n_r,\,n_t,\,p}\right]^T_{L_p\times 1}\\
\tilde{\mathbf{h}}_{k,\,n_r,\,n_t} &= \left[\tilde{h}_{k,\,0,\,n_r,\,n_t} \;\cdots\; \tilde{h}_{k,\,L_{hr}-1,\,n_r,\,n_t}\right]^T_{L_{hr}\times 1}\\
\tilde{\mathbf{s}}_5 &= \begin{bmatrix} \tilde{s}_{5,\,L_{hr}-1} & \cdots & \tilde{s}_{5,\,0}\\ \vdots & \cdots & \vdots \\ \tilde{s}_{5,\,L_p+L_{hr}-2} & \cdots & \tilde{s}_{5,\,L_p-1}\end{bmatrix}_{L_p\times L_{hr}}.
\end{aligned} \tag{33}$$

Observe that $\tilde{\mathbf{s}}_5$ is independent of m_1 and due to the relations in Eqs. (10), (15), and (21), we have

$$\tilde{\mathbf{s}}_5^H\tilde{\mathbf{s}}_5 = \frac{8L_p}{L_d}\mathbf{I}_{L_{hr}}. \tag{34}$$

where $\mathbf{I}_{L_{hr}}$ is an $L_{hr} \times L_{hr}$ identity matrix. The estimate of the channel is [21, 36, 39]

$$\hat{\mathbf{h}}_{k, n_r, n_t} = \left(\tilde{\mathbf{s}}_5^H \tilde{\mathbf{s}}_5\right)^{-1} \tilde{\mathbf{s}}_5^H \tilde{\mathbf{r}}_{k, m_1, n_r, n_t, p}. \tag{35}$$

To see the effect of noise on the channel estimate in Eq. (35), consider

$$\tilde{\mathbf{u}} = \left(\tilde{\mathbf{s}}_5^H \tilde{\mathbf{s}}_5\right)^{-1} \tilde{\mathbf{s}}_5^H \tilde{\mathbf{w}}_{k, m_1, n_r, n_t, p}. \tag{36}$$

It can be shown that [21, 39]

$$E\left[\tilde{\mathbf{u}}\tilde{\mathbf{u}}^H\right] = \frac{\sigma_w^2 L_d}{4L_p} \mathbf{I}_{L_{hr}} \stackrel{\Delta}{=} 2\sigma_u^2 \mathbf{I}_{L_{hr}}. \tag{37}$$

3.3 Noise variance estimation

The noise variance per dimension is estimated as

$$\begin{aligned} \hat{\sigma}_w^2 = \frac{1}{2L_p N^2 N_{rt}} \sum_{k=1}^{N_{rt}} \sum_{n_t=1}^{N} \sum_{n_r=1}^{N} \left(\tilde{\mathbf{r}}_{k, m_1, n_r, n_t, p} - \tilde{\mathbf{s}}_5 \hat{\mathbf{h}}_{k, n_r, n_t}\right)^H \\ \left(\tilde{\mathbf{r}}_{k, m_1, n_r, n_t, p} - \tilde{\mathbf{s}}_5 \hat{\mathbf{h}}_{k, n_r, n_t}\right). \end{aligned} \tag{38}$$

3.4 Post-FFT operations

In this section, we assume that the residual frequency offset given by

$$\omega_r = \omega_0 - \hat{\omega}_0 \tag{39}$$

is such that

$$\omega_r L_d < 0.1 \quad \text{radians} \tag{40}$$

so that the effect of inter carrier interference (ICI) is negligible. Let

$$m_2 = m_1 + N\left(L_p + L_{cp}\right) \tag{41}$$

where m_1 is defined in Eq. (31). Note that m_2 is the starting point of the data phase. Define the FFT input in the time domain for the kth retransmission and receive antenna n_r as

$$\tilde{\mathbf{r}}_{k, m_2, n_r, d} = \left[\tilde{r}_{k, m_2, n_r, d} \quad \dots \quad \tilde{r}_{k, m_2+L_d-1, n_r, d}\right]_{L_d \times 1}^T \tag{42}$$

where we have followed the notation in Eq. (16). The L_d-point FFT of Eq. (42) is

$$\tilde{\mathbf{R}}_{k, n_r, d} = \left[\tilde{R}_{k, 0, n_r, d} \quad \dots \quad \tilde{R}_{k, L_d-1, n_r, d}\right]_{L_d \times 1}^T \tag{43}$$

where $\tilde{R}_{k, i, n_r, d}$ is given by Eq. (18). Construct a matrix:

$$\tilde{\mathbf{R}}_{k, i, d} = \left[\tilde{R}_{k, i, 1, d} \quad \dots \quad \tilde{R}_{k, i, N, d}\right]_{N \times 1} \tag{44}$$

for $0 \le i \le L_d - 1$. Note that from Eq. (18)

$$\tilde{\mathbf{R}}_{k,i,d} = \tilde{\mathbf{H}}_{k,i}\mathbf{S}_{3,i} + \tilde{\mathbf{W}}_{k,i,d} \tag{45}$$

where

$$\tilde{\mathbf{H}}_{k,i} = \begin{bmatrix} \tilde{H}_{k,i,1,1} & \dots & \tilde{H}_{k,i,1,N} \\ \vdots & \dots & \vdots \\ \tilde{H}_{k,i,N,1} & \dots & \tilde{H}_{k,i,N,N} \end{bmatrix}_{N\times N}$$
$$\mathbf{S}_{3,i} = \left[S_{3,i,1} \quad \dots \quad S_{3,i,N}\right]^T_{N\times 1} \tag{46}$$
$$\tilde{\mathbf{W}}_{k,i,d} = \left[\tilde{W}_{k,i,1,d} \quad \dots \quad \tilde{W}_{k,i,N,d}\right]^T_{N\times 1}.$$

which is similar to Eq. (1) in [34]. Let

$$\tilde{\mathbf{Y}}_{k,i} = \hat{\mathbf{H}}^H_{k,i}\tilde{\mathbf{R}}_{k,i,d} = \hat{\mathbf{H}}^H_{k,i}\tilde{\mathbf{H}}_{k,i}\mathbf{S}_{3,i} + \hat{\mathbf{H}}^H_{k,i}\tilde{\mathbf{W}}_{k,i,d} \tag{47}$$

where $\hat{\mathbf{H}}_{k,i}$ is constructed from the L_d-point FFT of $\hat{\mathbf{h}}_{k,n_r,n_t}$ in Eq. (35) and $\hat{\mathbf{Y}}_{k,i}$ is similar to $\tilde{\mathbf{Y}}_k$ in Eq. (6) of [34]. The analysis when

$$\hat{\mathbf{H}}_{k,i} = \tilde{\mathbf{H}}_{k,i} \tag{48}$$

is given in [34]. Let

$$\tilde{\mathbf{Y}}_i = \frac{1}{N_{rt}}\sum_{k=1}^{N_{rt}} \tilde{\mathbf{Y}}_{k,i} \qquad \text{for } 0 \le i \le L_d - 1. \tag{49}$$

Note that $\tilde{\mathbf{Y}}_i$ is an $N \times 1$ matrix, whose n_t^{th} element $\tilde{Y}_{i,n_t}$ is a noisy version of S_{3,i,n_t} in Eq. (18). The matrix

$$\tilde{\mathbf{Y}}_{n_t} = \left[\tilde{Y}_{0,n_t} \dots \tilde{Y}_{L_d-1,n_t}\right]^T_{L_d\times 1} \tag{50}$$

constructed from the elements of $\tilde{\mathbf{Y}}_i$ in Eq. (49) is fed to the turbo decoder. The forward (α) backward (β) recursions for decoder 1 of the turbo code is given by Eqs. (28) and (31) in [34]. The term $\gamma_{1,i,m,n}$ in Eq. (30) of [34] should be replaced by

$$\gamma_{1,i,m,n,n_t} = \exp\left[-\frac{\left|\tilde{Y}_{i,n_t} - F_{i,n_t}S_{m,n}\right|^2}{2\sigma_U^2}\right] \tag{51}$$

where $\tilde{Y}_{i,n_t}$ is an element of $\tilde{\mathbf{Y}}_{n_t}$ in Eq. (50) and n_t is an odd integer. The term σ_U^2 in Eq. (51) is given by Eq. (22) in [34] which is repeated here for convenience:

$$\sigma_U^2 = \frac{1}{N_{rt}}\left(4N\sigma_H^2\sigma_W^2 + 8N(N-1)\sigma_H^4\right) \tag{52}$$

with

$$\sigma_W^2 = L_d\hat{\sigma}_w^2$$
$$\sigma_H^2 = \frac{1}{2N^2N_{rt}L_d}\sum_{k=1}^{N_{rt}}\sum_{i=0}^{L_d-1}\sum_{n_r=1}^{N}\sum_{n_t=1}^{N}\left|\hat{H}_{k,i,n_r,n_t}\right|^2 \tag{53}$$

where $\hat{\sigma}_w^2$ is given by Eq. (38) and $\hat{H}_{k,i,n_r,n_t}$ is obtained by taking the L_d-point FFT of (35). The term F_{i,n_t} in Eq. (51) is given by

$$F_{i,n_t} = \frac{1}{N_{rt}} \sum_{k=1}^{N_{rt}} F_{k,i,n_t} \tag{54}$$

where

$$F_{k,i,n_t} = \sum_{n_r=1}^{N} \left|\hat{H}_{k,i,n_r,n_t}\right|^2. \tag{55}$$

The extrinsic information from decoder 1 to decoder 2 is computed using Eqs. (32) and (33) of [34], with $\gamma_{1,i,n,\rho^+(n)}$ replaced by $\gamma_{1,i,n,\rho^+(n),n_t}$. The equations for decoder 2 are similar, except that $\gamma_{2,i,m,n}$ in Eq. (34) of [34] should be replaced by

$$\gamma_{2,i,m,n,n_t+1} = \exp\left[-\frac{\left|\tilde{Y}_{i,n_t+1} - F_{i,n_t+1} S_{m,n}\right|^2}{2\sigma_U^2}\right] \tag{56}$$

where again n_t is an odd integer.

3.5 Throughput and spectral efficiency

Recall from **Figure 3(a)** that during the preamble phase, only one transmit antenna is active at a time, whereas during the data phase, all the transmit antennas are simultaneously active. Thus the throughput can be defined as [36, 37].

$$\mathcal{T} = \frac{N L_d/2}{N_{rt}\left[N\left(L_p + L_{cp}\right) + L_d + L_{cp}\right]}. \tag{57}$$

The numerator of Eq. (57) denotes the total number of data bits transmitted, and the denominator represents the total number of QPSK symbol *durations* over N_{rt} retransmissions. The symbol rate during the preamble phase and data phase is the same. In the data phase, we are transmitting coded QPSK, that is, in each data bit duration, two coded QPSK symbols are sent simultaneously from two transmit antennas (see **Figure 3(b)**). Thus, during the data phase, each transmit antenna sends half a bit of information in each transmission. Therefore, the spectral efficiency is

Simulation parameters		Throughput $\mathcal{T}$
L_p = 512		
L_d = 1024	N = 4	32.38%
L_{cp} = 18 N_{rt} = 2	N = 8	38.77%
L_p = 4096		
L_d = 8192	N = 4	33.21%
L_{cp} = 18 N_{rt} = 2	N = 8	39.84%

Table 1.
Throughput for various simulation parameters.

$$\mathscr{S} = N/(2N_{rt}) \quad \text{bits per transmission.} \tag{58}$$

The throughput for various simulation parameters is given in **Table 1**. Observe that when $L_p = L_d/2$, $L_{cp} \ll L_d$, and $N \gg 1$, $\mathscr{T} \to 1/N_{rt}$. In this work, we have used a rate-1/2 turbo code, that is, each data bit generates two coded QPSK symbols. The throughput can be doubled by using a rate-1 turbo code, obtained by puncturing.

4. Simulation results

The simulation parameters are given in **Table 2**. A "run" in **Table 2** is defined as transmitting and receiving the frame in **Figure 3(a)** over N_{rt} retransmissions. The generating matrix of each of the constituent encoders of the turbo code is given by Eq. (49) in [21]. A question might arise: how does $N = 4, 8$ correspond to a massive MIMO system, whereas in [34] N was as large as 512? The answer is in [34], an ideal massive MIMO was considered, wherein the channel, timing, and carrier frequency offset were assumed to be known, whereas in this work, the channel, timing, and carrier frequency offset are estimated. The estimation complexity and memory requirement increase as N^2, for an $N \times N$ MIMO system. For example, the memory requirement of Eq. (27) when the number of frequency bins $B_1 = 1024$ [21], preamble length $L_p = 4096$, cyclic prefix length $L_{cp} = 18$, channel length $L_h = 10$, $N = 8$ transmit and receive antennas, and $N_{rt} = 4$ retransmissions is

$$\begin{aligned} \text{memory requirement} &= \left(L_p + L_{cp} + L_h - 1\right)(B_1 + 1)N^2 N_{rt} \\ &= 1081875200 \end{aligned} \tag{59}$$

double precision values. In fact Eq. (27) is implemented using multidimensional arrays in Scilab, instead of using for loops. Note that from Eq. (8), the length of the received signal during the preamble phase is $L_p + L_{cp} + L_h - 1$. If for loops are used, the memory requirement would be

$$\begin{aligned} \text{memory requirement} &= \left(L_p + L_{cp} + L_h - 1\right)(B_1 + 1) \\ &= 4226075 \end{aligned} \tag{60}$$

double precision values, which is much less than Eq. (59); however the simulations would run much slower. Does this mean that we cannot go higher than an

Parameter	Value
L_p	512, 4096
L_d	1024, 8192
L_h	10
L_{hr}	19
L_{cp}	18
N	4, 8
N_{rt}	1, 2, 4
B_1 [21]	64, 1024
Runs	10^4, 10^3

Table 2.
Simulation parameters.

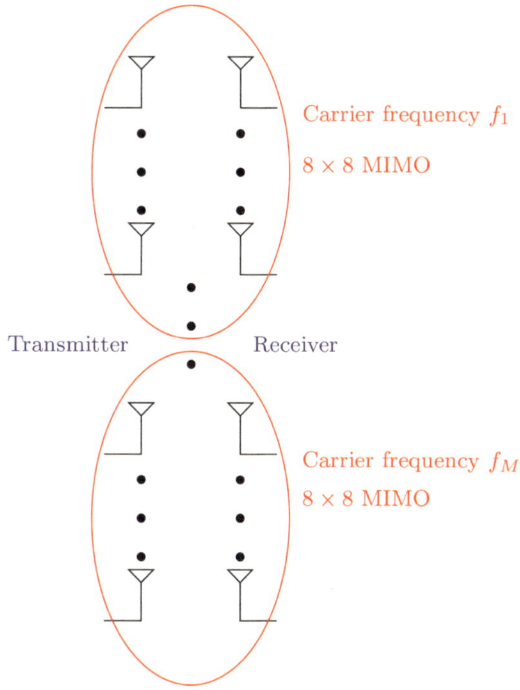

Figure 4.
Massive MIMO using multiple carrier frequencies.

8 × 8 MIMO system? The answer is no. The solution lies in using multiple carrier frequencies as illustrated in **Figure 4**. Observe that with 8 × 8 MIMO and M carrier frequencies, we get an overall $8M \times 8M$ MIMO system. The bit error rate results for a 4 × 4 MIMO system are shown in **Figure 5**. The bit error rate results for an 8 × 8 MIMO system are shown in **Figure 6**. The following observations can be made from **Figure 5**:

1. There is only 0.75 dB difference in performance between the ideal (id) and estimated (est) receiver, for $L_d = 1024$, $N_{rt} = 2$, and bit error rate equal to 10^{-4}. On the other hand, there is hardly any performance difference between the ideal and estimated receiver for $L_d = 8192$. This is because the noise

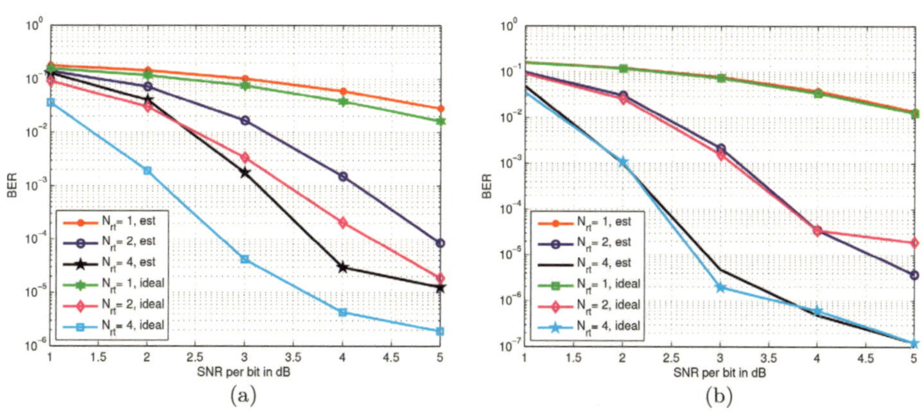

Figure 5.
Bit error rate results for a 4 × 4 MIMO system. (a) $L_d = 1024$ and $L_p = 512$. (b) $L_d = 8192$ and $L_p = 4096$.

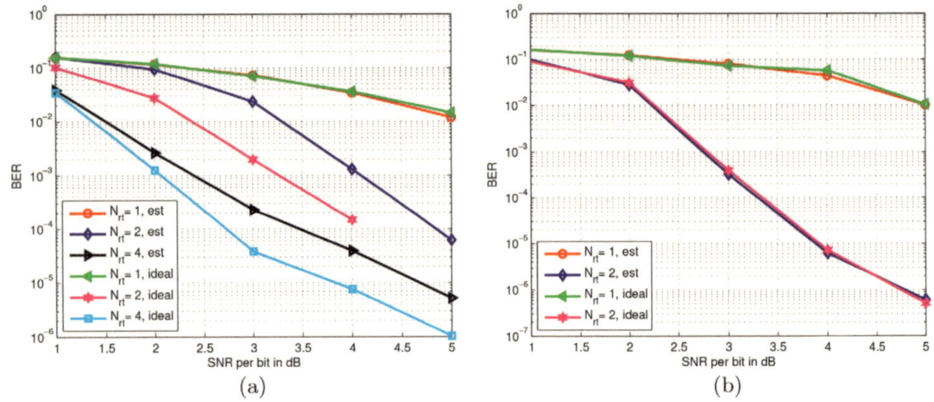

Figure 6.
Bit error rate results for an 8×8 *MIMO system. (a)* $L_d = 1024$ *and* $L_p = 512$*. (b)* $L_d = 8192$ *and* $L_p = 4096$.

variance (σ_w^2) decreases with increasing L_p, L_d, for a given average SNR per bit, as shown by Eq. (21).

2. There is only 0.5 dB improvement in performance for $L_d = 8192$ over $L_d = 1024$, at a BER of 10^{-4}.

3. There is a significant improvement in performance between $N_{rt} = 2$ and $N_{rt} = 4$, for both $L_d = 1024$ and $L_d = 8192$. On the other hand, there is no significant difference in the BER for $N_{rt} = 2$ and $N_{rt} = 4$ in [34]. The reason is because in this work, the channel $\tilde{H}_{k,i,n_r,n_t}$ in Eq. (18) is highly correlated over the subcarrier index i, since it is obtained by taking an L_d-point FFT of an L_h-tap channel (see Eq. (37) of [36]). On the other hand, the channel $\tilde{H}_{k,i,j}$ in [34] is independent over all the indices k, i, and j, where k is the retransmission index, i denotes the receive antenna index, and j denotes the transmit antenna index. See also the discussion leading to Eq. (82) in [21].

4. There is not much BER performance difference between the 4×4 and 8×8 MIMO systems. A 4×4 MIMO is computationally less complex than 8×8; however the 4×4 requires twice the number of carrier frequencies to achieve the same spectral efficiency as 8×8, for the same number of retransmissions.

5. The BER performance of the 8×8 MIMO system with $N_{rt} = 4$ and $L_d = 8192$ could not be simulated due to the large amount of memory involved (see Eq. (59)) and Scilab limitations.

5. Conclusions

This work describes the discrete-time algorithms for the implementation of a massive MIMO system. Due to the implementation complexity considerations, more than one carrier frequency is required to obtain a truly single-user massive MIMO system. Each carrier frequency needs to be associated with an 8×8 or 4×4 MIMO subsystem. The average SNR per bit has been used as a performance measure, which has not been done earlier in the literature. Perhaps the channel can also be estimated using Eq. (27), instead of using Eq. (35). This needs investigation.

Acknowledgements

The authors would like to thank the high-performance computing (HPC) facility at IIT Kanpur for running the computer simulations.

Author details

K. Vasudevan*, Shivani Singh and A. Phani Kumar Reddy
Department of Electrical Engineering, Indian Institute of Technology, Kanpur, India

*Address all correspondence to: vasu@iitk.ac.in

References

[1] Pi Z, Khan F. An introduction to millimeter-wave mobile broadband systems. IEEE Communications Magazine. 2011;**49**(6):101-107

[2] Rusek F, Persson D, Lau BK, Larsson EG, Marzetta TL, Edfors O, et al. Scaling up MIMO: Opportunities and challenges with very large arrays. IEEE Signal Processing Magazine. 2013; **30**(1):40-60

[3] Hoydis J, ten Brink S, Debbah M. Massive MIMO in the UL/DL of cellular networks: How many antennas do we need? IEEE Journal on Selected Areas in Communications. 2013;**31**(2): 160-171

[4] Rappaport TS, Sun S, Mayzus R, Zhao H, Azar Y, Wang K, et al. Millimeter wave mobile communications for 5G cellular: It will work! IEEE Access. 2013;**1**:335-349

[5] Chih-Lin I, Rowell C, Han S, Xu Z, Li G, Pan Z. Toward green and soft: A 5G perspective. IEEE Communications Magazine. 2014;**52**(2):66-73

[6] Larsson EG, Edfors O, Tufvesson F, Marzetta TL. Massive MIMO for next generation wireless systems. IEEE Communications Magazine. 2014;**52**(2): 186-195

[7] Andrews JG, Buzzi S, Choi W, Hanly SV, Lozano A, Soong ACK, et al. What will 5G be? IEEE Journal on Selected Areas in Communications. 2014;**32**(6): 1065-1082

[8] Rappaport TS, Roh W, Cheun K. Mobile's millimeter-wave makeover. IEEE Spectrum. 2014;**51**(9):34-58

[9] Marzetta TL. Massive MIMO: An introduction. Bell Labs Technical Journal. 2015;**20**:11-22

[10] Galinina O, Pyattaev A, Andreev S, Dohler M, Koucheryavy Y. 5G multi-rat LTE-WiFi ultra-dense small cells: Performance dynamics, architecture, and trends. IEEE Journal on Selected Areas in Communications. 2015;**33**(6): 1224-1240

[11] Bjrnson E, Larsson EG, Debbah M. Massive MIMO for maximal spectral efficiency: How many users and pilots should be allocated? IEEE Transactions on Wireless Communications. 2016; **15**(2):1293-1308

[12] Chih-Lin I, Han S, Xu Z, Wang S, Sun Q, Chen Y. New paradigm of 5G wireless internet. IEEE Journal on Selected Areas in Communications. 2016;**34**(3):474-482

[13] Agiwal M, Roy A, Saxena N. Next generation 5G wireless networks: A comprehensive survey. IEEE Communication Surveys and Tutorials. 2016;**18**(3):1617-1655

[14] Wong KL, Tsai CY, Lu JY, Chian DM, Li WY. Compact eight MIMO antennas for 5G smartphones and their MIMO capacity verification. In: 2016 URSI Asia-Pacific Radio Science Conference (URSI AP-RASC); 2016. pp. 1054-1056

[15] Buzzi S, D'Andrea C. Doubly massive mmWave MIMO systems: Using very large antenna arrays at both transmitter and receiver. In: 2016 IEEE Global Communications Conference (GLOBECOM); 2016. pp. 1-6

[16] Mesleh RY, Haas H, Sinanovic S, Ahn CW, Yun S. Spatial modulation. IEEE Transactions on Vehicular Technology. 2008;**57**(4):2228-2241

[17] Renzo MD, Haas H, Grant PM. Spatial modulation for multiple-antenna

wireless systems: A survey. IEEE Communications Magazine. 2011; **49**(12):182-191

[18] Renzo MD, Haas H, Ghrayeb A, Sugiura S, Hanzo L. Spatial modulation for generalized MIMO: Challenges, opportunities, and implementation. Proceedings of the IEEE. 2014;**102**(1): 56-103

[19] Cao Y, Ohtsuki T, Jiang X. Precoding aided generalized spatial modulation with an iterative greedy algorithm. IEEE Access. 2018;**6**: 72449-72457

[20] Pan Z, Luo J, Lei J, Wen L, Tang C. Uplink spatial modulation SCMA system. IEEE Communications Letters. 2019;**23.1**:184-187

[21] Vasudevan K. Near capacity signaling over fading channels using coherent turbo coded OFDM and massive MIMO. International Journal on Advances in Telecommunications. 2017; **10**(1 & 2):22-37

[22] Godara LC. Application of antenna arrays to mobile communications. II. Beam-forming and direction-of-arrival considerations. Proceedings of the IEEE. 1997;**85**(8):1195-1245

[23] Roh W, Seol JY, Park J, Lee B, Lee J, Kim Y, et al. Millimeter-wave beamforming as an enabling technology for 5G cellular communications: Theoretical feasibility and prototype results. IEEE Communications Magazine. 2014;**52**(2):106-113

[24] Vikalo H, Hassibi B, Kailath T. Iterative decoding for mimo channels via modified sphere decoding. IEEE Transactions on Wireless Communications. 2004;**3**(6):2299-2311

[25] Jalden J, Ottersten B. On the complexity of sphere decoding in digital communications. IEEE Transactions on Signal Processing. 2005;**53**(4): 1474-1484

[26] Burg A, Borgmann M, Wenk M, Zellweger M, Fichtner W, Bolcskei H. VLSI implementation of MIMO detection using the sphere decoding algorithm. IEEE Journal of Solid-State Circuits. 2005;**40**(7):1566-1577

[27] Hassibi B, Vikalo H. On the sphere-decoding algorithm. I. Expected complexity. IEEE Transactions on Signal Processing. 2005;**53**(8): 2806-2818

[28] Studer C, Burg A, Bolcskei H. Soft-output sphere decoding: Algorithms and VLSI implementation. IEEE Journal on Selected Areas in Communications. 2008;**26**(2):290-300

[29] Barbero LG, Thompson JS. Fixing the complexity of the sphere decoder for MIMO detection. IEEE Transactions on Wireless Communications. 2008;**7**(6): 2131-2142

[30] Nguyen TVH, Sugiura S, Lee K. Low-complexity sphere search-based adaptive spatial modulation. IEEE Transactions on Vehicular Technology. 2018;**67**(8):7836-7840

[31] Li L, Wen J, Tang X, Tellambura C. Modified sphere decoding for sparse code multiple access. IEEE Communications Letters. 2018;**22**(8): 1544-1547

[32] Karamanakos P, Geyer T, Aguilera RP. Long-horizon direct model predictive control: Modified sphere decoding for transient operation. IEEE Transactions on Industry Applications. 2018;**54**(6):6060-6070

[33] Tran TQ, Sugiura S, Lee K. Ordering- and partitioning-aided sphere decoding for generalized spatial

modulation. IEEE Transactions on Vehicular Technology. 2018;**67**(10): 10087-10091

[34] Vasudevan K, Madhu K, Singh S. Data detection in single user massive MIMO using re-transmissions. The Open Signal Processing Journal. 2019;**6**: 15-26. arXiv preprint arXiv:1811.11369. 2019. DOI: 10.2174/1876825301906 010015

[35] Vasudevan K. Digital Communications and Signal Processing. 2nd ed. (CDROM included). Hyderabad: Universities Press (India); 2010. www.universitiespress.com

[36] Vasudevan K. Coherent detection of turbo coded OFDM signals transmitted through frequency selective Rayleigh fading channels. In: 2013 IEEE International Conference on Signal Processing, Computing and Control (ISPCC); 2013. pp. 1-6

[37] Vasudevan, K. Coherent detection of turbo-coded OFDM signals transmitted through frequency selective Rayleigh fading channels with receiver diversity and increased throughput. Wireless Personal Communications. 2015;**82**(3):1623-1642. DOI: 10.1007/s11277-015-2303-8

[38] Vasudevan K. Coherent detection of turbo-coded OFDM signals transmitted through frequency selective Rayleigh fading channels with receiver diversity and increased throughput. CoRR. Vol. abs/1511.00776; 2015. Available Online: http://arxiv.org/abs/1511.00776

[39] Vasudevan K. Coherent turbo coded MIMO OFDM. In: ICWMC 2016 - The 12th International Conference on Wireless and Mobile Communications; 2016. pp. 91-99

[40] Veludandi VK, Vasudevan K. Linear prediction-based detection of serially concatenated DQPSK in SIMO-OFDM. Wireless Personal Communications. 2017;**97**(3):4789-4811. DOI: 10.1007/s11277-017-4751-9

[41] Mishra HB, Vasudevan K. Design of superimposed training sequence for spatially correlated multiple-input multiple-output channels under interference-limited environments. IET Communications. 2015;**9**:1259-1268(9)

[42] Minn H, Bhargava VK, Letaief KB. A robust timing and frequency synchronization for OFDM systems. IEEE Transactions on Wireless Communications. 2003;**2**(4):822-839

[43] Tarokh V, Jafarkhani H. On the computation and reduction of the peak-to-average power ratio in multicarrier communications. IEEE Transactions on Communications. 2000;**48**(1):37-44

[44] Han SH, Lee JH. An overview of peak-to-average power ratio reduction techniques for multicarrier transmission. IEEE Wireless Communications. 2005;**12**(2):56-65

[45] Jiang T, Zhu G. Complement block coding for reduction in peak-to-average power ratio of OFDM signals. IEEE Communications Magazine. 2005;**43**(9): S17-S22

[46] Jiang T, Wu Y. An overview: Peak-to-average power ratio reduction techniques for OFDM signals. IEEE Transactions on Broadcasting. 2008; **54**(2):257-268

[47] Rahmatallah Y, Mohan S. Peak-to-average power ratio reduction in OFDM systems: A survey and taxonomy. IEEE Communication Surveys and Tutorials. 2013;**15**(4):1567-1592

[48] Wunder G, Fischer RFH, Boche H, Litsyn S, No J. The PAPR problem in OFDM transmission: New directions for

a long-lasting problem. IEEE Signal Processing Magazine. 2013;**30**(6): 130-144

[49] Sharma E, Mishra HB, Vasudevan K, Budhiraja R. PAPR analysis of superimposed training based SISO/MIMO-OFDM systems with orthogonal affine precoder. Physical Communication. 2017;**25**(P1):239-248. DOI: 10.1016/j.phycom.2017.08.004

Section 4

Optic Applications

Chapter 5

Direct Sequence-Optical Code-Division Multiple Access (DS-OCDMA): Receiver Structures for Performance Improvement

Younes Zouine and Zhour Madini

Abstract

We present in this chapter, the performance study of a direct sequence-optical code-division multiple access (O-CDMA) link. In such systems, the main limitation is the multiple access interference (MAI). We investigate various schemes of receiver in the aim of improving the performances and mitigating MAI. Furthermore, we show the benefits of different techniques in regard to conventional ones. However, this system uses ultrashort light pulses that are sensible to the optical link parameters, especially the fiber chromatic dispersion. We have shown that when compensation dispersion devices are not deployed in the system, there is a trade-off between the limited dispersion effects and the MAI.

Keywords: PON, DS-OCDMA, multiple access interference, MAI, OOC, hard limiter, conventional correlation receiver, successive interference cancelation receiver (SIC), parallel interference cancelation receiver (PIC)

1. Introduction

The optical fiber offers a small footprint, a low attenuation, and especially a large bandwidth (estimated of the order of THz). However, the cost of a total redeployment of the optical fiber access network would be very important. In order to reduce these costs, it is possible to share the resource among several users, using a passive optical network (PON) type structure. In this case, it is necessary to set up multiple access techniques to differentiate the information associated with each user.

The two most widely used multiple access techniques for optical communications are time-division multiple access (TDMA) and wavelength-division multiple access (WDMA). These two techniques can constitute an economic brake, because the first requires the synchronization of all terminal equipment and the second one requires tunable wavelength filters to adapt to the desired wavelength. Another technique derived from radiofrequency systems has been envisaged for several decades for optical communications: code-division multiple access (CDMA) [1, 2].

In an OCDMA system, the manipulation of the signals can be considered either coherently or incoherently. In a coherent approach, the characteristics of the optical signal measured are amplitude and phase. This configuration requires having a local oscillator synchronized to the optical frequency on reception, which increases the cost of implementation. Since the light wave can be positive or negative, data spreading can be carried out using bipolar codes such as radiofrequency.

But most studies on optical CDMA are about inconsistent systems, much simpler and, therefore, less expensive. They are usually based on a modulation scheme called "intensity modulation-direct detection" (IM-DD), and it is the luminous intensity, positive quantity, which is the measured characteristic of the optical signal. Bipolar codes can no longer be used. Unipolar quasi-orthogonal codes are used.

This chapter is organized as follows: in Sections 1 and 2, a brief description on the state-of-the-art of the DS-OCDMA technique and its advantages over TDMA and WDMA are presented. In Section 3, a study of a DS-OCDMA system using a CCR and CCR with HL is reported. Section 4 is devoted to multiuser receivers such as SIC and PIC receivers. Section 5 is dedicated to the impact of chromatic dispersion on DS-OCDMA performances. Finally, conclusions are drawn in Section 6.

2. The optical CDMA network

Every user utilizes an amplitude shift keying (ASK) especially the on/off keying modulation to transmit all the binary data via a common optical fiber. The encoder impresses a sequence code upon the binary data (**Figure 1**). The sequence code is specific to each user, in order to be able to extract the data by correlation at the end receiver. For the data recovery, the received signal would be compared at first to the sequence code, and then to a threshold level at the comparator.

For low multiple access interference (MAI) and error probability, the chosen codes must have good correlation properties. In this chapter, we consider optical orthogonal codes (OOC) [1]. A class of codes is defined by (F, W, λ_a, λ_c) where F is the length of the sequences, W is the weight, and λ_a and λ_c the auto- and crosscorrelation constraints, respectively. The maximum number of users N in the OOC's class is defined as: $N(F, W, 1, 1) = \left\lfloor \frac{F-1}{W(W-1)} \right\rfloor$.

The code signature, $c_k(t)$, of the kth user is $c_k(t) = \sum_{j=-\infty}^{\infty} c_j^{(k)} P_{T_c}(t - jT_c)$ where $P_{Tc}(t)$ is a unit rectangular pulse with duration of one chip T_c and $\left\{c_j^{(k)}\right\} \in \{0,1\}$ is the jth value of the kth user spreading code. In general, in the case of $\lambda_a = \lambda_c = \lambda$, the number of users N is limited by the Johnson bound [1], given by the relation:

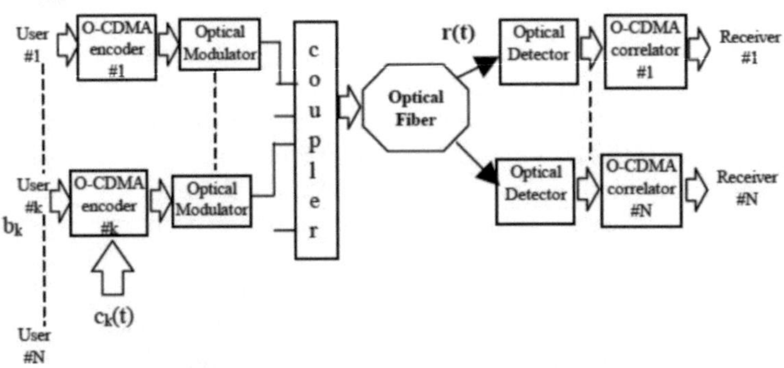

Figure 1.
Synoptic scheme of a DS-OCDMA system.

$$N(F, W, \lambda_a, \lambda_c) \leq \left\lfloor \frac{1}{W} \left\lfloor \frac{F-1}{W-1} \left\lfloor \frac{F-2}{W-2} \right\lfloor --- \left\lfloor \frac{F-\lambda}{W-\lambda} \right\rfloor --- \right\rfloor \right\rfloor \right\rfloor. \quad (1)$$

The N output signals are directed via optical devices, such as star couplers, toward all receivers along the optical fiber. The received signal, r(t), is the sum of signals transmitted by all active users:

$r(t) = \sum_{k=1}^{N} s_k(t - \tau_k)$ where s_k is the transmitted signal of kth user and τ_k is the delay of user k.

The transmitted signal s_k is given by the following equation: $s_k(t) = b_k(t)c_k(t)$ where

- $b_k(t) = \sum_{i=-\infty}^{\infty} b_i^{(k)} P_{T_b}(t - iT_b)$ represents the data of the kth user

- $b_i^{(k)}$ is the ith data bit of kth user, and $\left\{ b_i^{(k)} \right\} \in \{0,1\}$

- $P_{T_b}(t)$ is a rectangular pulse of duration T_b

$$r(t) = \sum_{k=1}^{N} \sum_{i=-\infty}^{\infty} \sum_{j=0}^{F-1} m_{i,j}^{k} P_{T_C}(t - iT_b - jT_c - \tau_k) \text{ where } m_{i,j}^{k} = C_j^k \cdot b_i^k. \quad (2)$$

3. Single-user detection

The receiver is a critical part, because according to its structure and its adequacy to the considered codes, it will condition the performances of the system. Among the main receivers envisaged for optical CDMA, one can distinguish different types:

- Single-user receivers, for which only the knowledge of the code of the desired user is necessary. For these receivers, the interference generated by the other users is not taken into account and is considered as noise. As this interference increases significantly with the number of active users, these receivers make many errors in a busy network.

- Multiuser receivers, for which knowledge of the codes of other users is required. These receivers are more complex than single-user receivers. They use knowledge of nondesired user codes to more reliably estimate the desired user. As a result, they allow better performance.

3.1 Conventional correlation receiver (CCR)

In a DS-OCDMA system using a CCR receiver, the spread spectrum is achieved by directly multiplying a signature code sequence with the data to be transmitted. It provides three steps:

- At first, in reception, the receiver multiplexes the received signal with the code of the desired user.

- In this step, the multiplier output signal is reformatted via an integrator in order to evaluate the total power per bit. The output signal presents the decision variable

- In the end, the decision variable will be compared to the value of the threshold S of the decision-making block in order to make the estimated data.

The block diagram of the desired user's conventional receiver is shown in **Figure 2**.

Assuming that the user #1 is the desired user, the decoding part of the DS-OCDMA system is performed by correlation. The received signal is multiplied by the code of the desired user:

$$r_{corr}(t) = r(t).c_1(t) = \left(\sum_{k=1}^{N} b_k(t).c_k(t)\right).c_1(t) \tag{3}$$

$$r_{corr}(t) = b_1(t).c_1(t) + \sum_{k=2}^{N} b_k(t).c_k(t).c_1(t). \tag{4}$$

At the output of the integration block, we will get the decision variable value $Z_i^{(1)}$

$$Z_i^{(1)} = \int_0^{T_b} r_{corr}(t)dt = \int_0^{T_b} \left(\sum_{k=1}^{N} b_k(t) \cdot c_k(t)\right) \cdot c_1(t)dt \tag{5}$$

$$\begin{aligned} Z_i^{(1)}(t) &= \underbrace{\int_0^{T_b} b_i^{(1)}(t) \cdot c_1(t)dt +}_{i} \underbrace{\int_0^{T_b} \left(\sum_{k=2}^{N} b_i^{(k)}(t) \cdot c_k(t)\right) \cdot c_1(t)dt}_{ii} \\ &= \int_0^{T_b} b_1(t) \cdot c_1(t)dt + \int_0^{T_b} I(t) \cdot c_1(t)dt \end{aligned} \tag{6}$$

where $I(t) = \sum_{k=2}^{N} b_i^{(k)}(t) \cdot c_k(t)$.

The second term of this expression represents the multiple access interference (MAI) term due to all the nondesired users. It depends on the number of active users N and the OOC's correlation properties.

Then, a comparison of the decision variable with the threshold level S has been done. Recording the comparison result, an estimation of the transmitted bit, $\hat{b}_1(t)$, is given. An error happened when the MAI term is greater than the threshold level S and the data $\hat{b}_1(t)$ are 0.

Analytical upper bound expression of the error probability has been demonstrated and presented by Salehi et al in [1]:

$$P_{es} \leq \frac{1}{2} \sum_{i=0}^{N-1} \binom{N-1}{i} \left(\frac{W^2}{2F}\right)^i \left(1 - \frac{W^2}{2F}\right)^{N-1-i} \tag{7}$$

Curves in **Figure 3** show that the performances are degraded as the number of users increases. If the threshold level value, S, is chosen close to W, the errors number decreases.

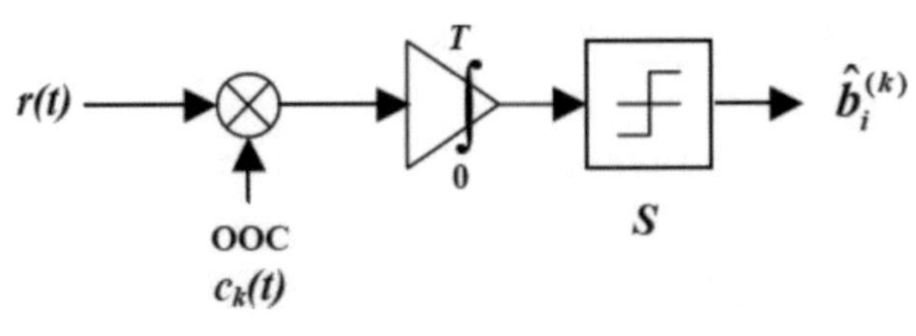

Figure 2.
The conventional correlation receiver (CCR).

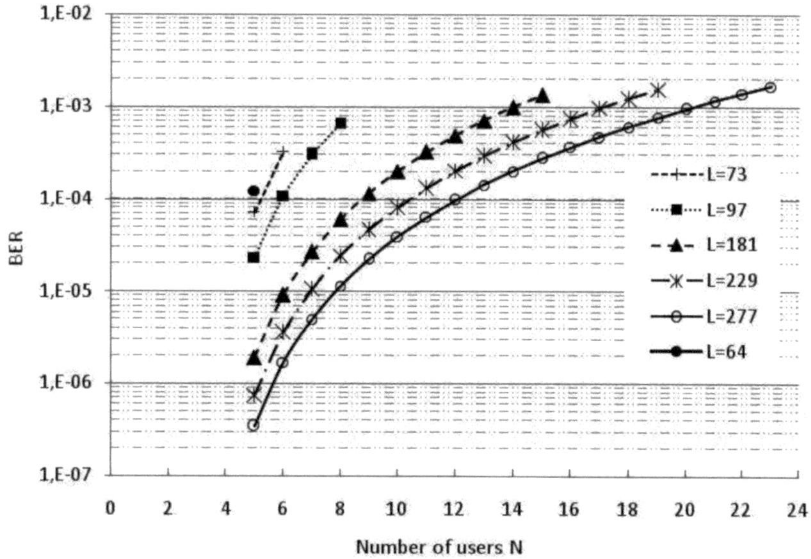

Figure 3.
BER as a function of the number of users for different lengths of the code.

3.2 Conventional correlation receiver with hard limiter (HL + CCR)

In order to limit the power of interfering users, an optical hard limiter (OHL) can be used. An ideal OHL function is defined as:

$$g(x) = \begin{cases} 0 & if \quad x < 1 \\ 1 & if \quad x \geq 1 \end{cases} \tag{8}$$

where 1 is the normalized optical power value of one chip.

So, if a received optical power x is bigger than or equal to 1, it will be clipped to 1. On the other hand, if a received optical power x is smaller than 1, it will be set to 0. Consequently, for S=W, an error can occur, if all the code chips are overlapped. Therefore, all other IAM configurations will be canceled. For the ideal chip synchronous case, the theoretical expression of the HL + CCR's error probability is given by [1, 3, 4].

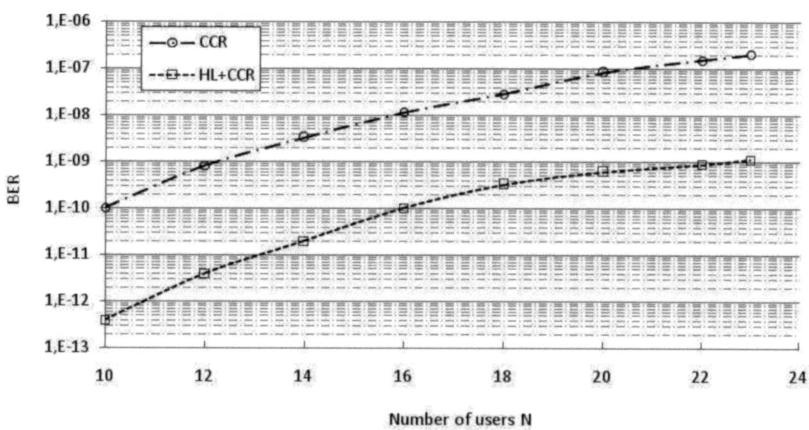

Figure 4.
Receiver's performances as a function of number of users.

$$P_{ehl} = \frac{1}{2}\binom{W}{S}\prod_{i=0}^{S-1}\left(1-\left(1-\frac{W^2}{2F}\right)^{N-1-i}\right) \tag{9}$$

Figure 4 shows the comparison between theoretical performances of the CCR and HL + CCR as a function of the active users number N, for F = 1000, W = 7, and S = 7. It is well verified that the HL + CCR provides better performance than the CCR.

4. Multiuser detection

In order to obtain better performances than those obtained by single-user detection, multiuser detection has been studied for OCDMA links [5–7]. Indeed, this type of detection, already used for wireless CDMA, has proved its effectiveness in reducing the impact of interference on performance.

The advantage of multiuser detection over single-user detection is the knowledge of nondesired user codes that allows finer evaluation of the interference present in the received signal. As a result, the data are better detected.

These detectors operate in two main stages:

1. estimating all or part of the interference present in the received signal.

2. detecting the desired user data after subtracting from received signal the estimated interference.

In the CDMA systems, we can distinguish between two types of multiuser detectors:

- Successive interference cancelation receiver (SIC).

- Parallel interference cancelation receiver (PIC).

4.1 Optical successive interference cancelation

The aim of the proposed method is the estimation of interference from interfering users. Once the IAM is determined, it is deducted from the received signal before detecting the desired user. The optical successive interference cancelation structure is shown in **Figure 5**. We assume that the desired user is the user n°# 1.

The optical-SIC receiver has the knowledge of all the active users' codes patterns, and we assume that all the users have the same transmitting energy. So, there are no strongest interfering signals [8].

The first step provides the estimation $\hat{b}_i^{(N)}$, for example, of the data of the Nth nondesired user by application of the traditional correlation method.

Then, the transmitting signal of the Nth user is reproduced and removed from the received signal r(t). We call the signal received after the subtraction, $s_1(t)$, which referred to output signal of the first cancelation stage (Cs = 1):

$$\begin{aligned} s_1(t) &= r(t) - \hat{b}_i^{(N)} c_N(t) \\ &= \hat{b}_i^{(1)} c_1(t) + \sum_{j=2}^{N-1} \hat{b}_i^{(j)} c_j(t) + \left(b_i^{(N)} - \hat{b}_i^{(N)}\right) c_N(t) \end{aligned} \tag{10}$$

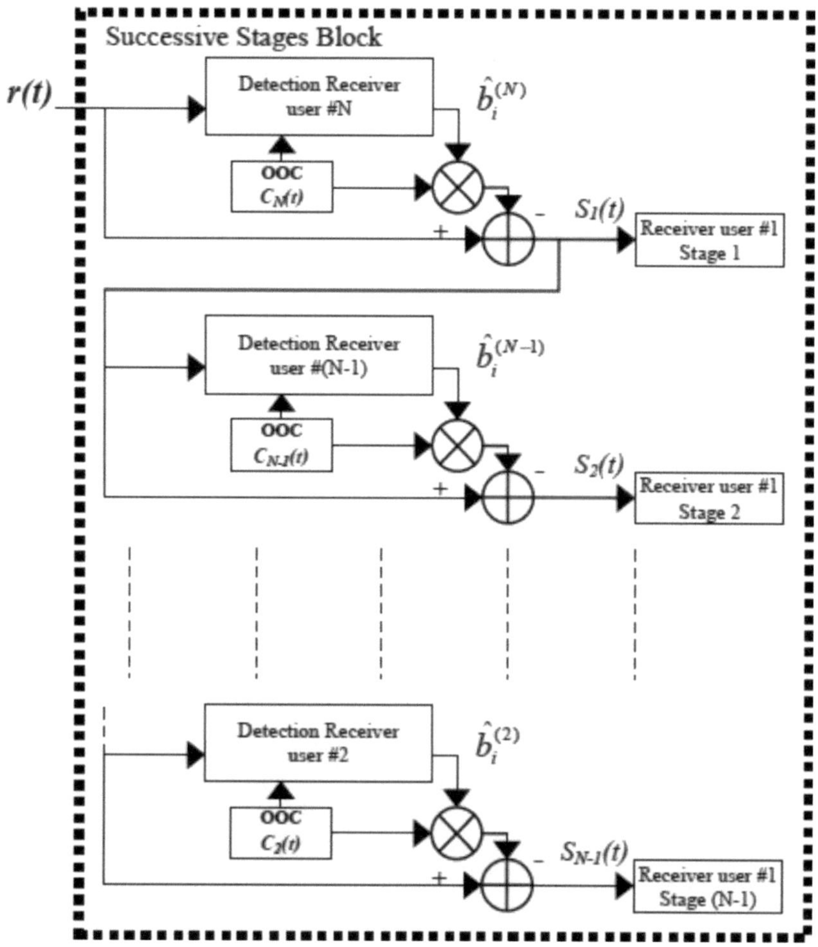

Figure 5.
Principle of the optical-SIC receiver.

The next step is the detection and estimation of the desired user data, for example, user 1, by conventional correlation method. The suppression of the undesired users could take until N−1 stages. At the output of the successive stages block, we obtain such signal before the conventional receiver for the user 1:

$$\begin{aligned} s_{N-1}(t) &= s_{N-2}(t) - \hat{b}_i^{(N)} c_N(t) \\ &= s_{N-3}(t) - \hat{b}_i^{(N)} c_N(t) - \hat{b}_i^{(N-1)} c_{N-1}(t) \\ &\dots \\ &= r(t) - \sum_{j=2}^{N} \hat{b}_i^{(j)} c_j(t) \\ &= \sum_{j=1}^{N} b_i^{(j)} c_j(t) - \sum_{j=2}^{N} \hat{b}_i^{(j)} c_j(t) \\ &= b_i^{(1)} c_1(t) + \sum_{j=2}^{N} \left(b_i^{(j)} - \hat{b}_i^{(j)} \right) \cdot c_j(t) \end{aligned} \tag{11}$$

4.1.1 First cancelation stage

We consider that noise perturbation is less significant than interference contribution, so the performances are linked only to MAI.

The decision variable for the desired user #1 is $Z_i^{(1)} = Wb_i^{(1)} + I_1 + A_1$ where:

- $I_1 = \sum_{k=3}^{N} \int_0^{T_b} b_i^{(k)} c_k(t) c_1(t) dt$ is The MAI at the output of the other users
- $A_1 = \left(b_i^{(N)} - \hat{b}_i^{(N)}\right) \int_0^{T_b} c_N(t) c_1(t) dt$ term linked to the cancelation of the user #N.

The cancelation term A1 can only take two values 0 and −1. Indeed, as we have conventional receivers on this stage, there are only errors for the data $b_i^{(N)}$ equal to 0, so:

$$A_1 = \begin{cases} 0 & si\ (b_i^{(N)} = \hat{b}_i^{(N)}\ \text{(if no error in detection)} \\ -1 & \text{if error in detection} \end{cases}.$$

The error probability is

$$P_{e_1} = \frac{1}{2} P_{e_{10}} + \frac{1}{2} P_{e_{11}}. \tag{12}$$

where $P_{e_{10}} = prob\left(\hat{b}_i^{(1)} = 1/b_i^{(1)} = 0\right)$ and $P_{e_{11}} = prob\left(\hat{b}_i^{(1)} = 0/b_i^{(1)} = 1\right)$

$$\begin{aligned} P_{e_{11}} &= prob\left(Z_i^{(1)} < S_1/b_i^{(1)} = 1\right) \\ &= prob\left(W + I_1 + A_1 < S_1/b_i^{(1)} = 1\right) \end{aligned} \tag{13}$$

$$\begin{aligned} P_{e_{11}} =& prob\left(A_1 = -1/b_i^{(1)} = 1\right) \cdot prob\left(I_1 < S_1 - W + 1/b_i^{(1)} = 1\right) \\ &+ prob\left(A_1 = 0/b_i^{(1)} = 1\right) \cdot prob\left(I_1 < S_1 - W/b_i^{(1)} = 1\right) \end{aligned} \tag{14}$$

$$prob\left(I_1 < S_1 - W + 1/b_i^{(1)} = 1\right) = \sum_{i=0}^{S_1 - W} C_{N-2}^{i} \left(\frac{W^2}{2L}\right)^i \left(1 - \frac{W^2}{2L}\right)^{N-2-i}. \tag{15}$$

We define the function

$$f(a, b, k) = \sum_{i=a}^{b} C_k^i \left(\frac{W^2}{2L}\right)^i \left(1 - \frac{W^2}{2L}\right)^{k-i}. \tag{16}$$

f(a,b,k) = 0 si a > b

then $prob\left(I_1 < S_1 - W + 1/b_i^{(1)} = 1\right) = f(0, S_1 - W, N - 2)$.

Similarly, on the one hand, we have

$prob\left(I_1 < S_1 - W/b_i^{(1)} = 1\right) = f(0, S_1 - W - 1, N - 2)$.

On the other hand, we have

$$
\begin{aligned}
prob\left(A_1 = -1/b_i^{(1)} = 1\right) &= prob\left(b_i^{(N)} - \hat{b}_i^{(N)} = -1/b_i^{(1)} = 1\right) \cdot prob\left(\int_0^{T_b} c_N(t)c_1(t)dt = 1\right) \\
&= \frac{W^2}{L} \cdot \frac{1}{2} prob\left(\hat{b}_i^{(N)} = 1/b_i^{(N)} = 0/b_i^{(1)} = 1\right) \\
&= \frac{1}{2}\frac{W^2}{L} prob\left(Z_i^{(N)} \geq S_N/b_i^{(1)} = 1 \ et \ b_i^{(N)} = 0\right) \\
&= \frac{1}{2}\frac{W^2}{L} f(S_N - 1, N-2, N-2) \\
prob\left(A_1 = 0/b_i^{(1)} = 1\right) &= 1 - prob\left(A_1 = -1/b_i^{(1)} = 1\right) \\
&= 1 - \frac{1}{2}\frac{W^2}{L} f(S_N - 1, N-2, N-2)
\end{aligned} \quad ; \tag{17}
$$

hence

$$
\begin{aligned}
P_{e_{11}} &= \frac{1}{2}\frac{W^2}{L} f(S_N - 1, N-2, N-2) \cdot f(0, S_1 - W, N-2) \\
&+ \left(1 - \frac{1}{2}\frac{W^2}{L} f(S_N - 1, N-2, N-2)\right) \cdot f(0, S_1 - W - 1, N-2) \, . \\
&= \frac{1}{2}\frac{W^2}{L} f(S_N - 1, N-2, N-2) \cdot f(0, S_1 - W, N-2)
\end{aligned} \tag{18}
$$

In the same way, we calculate the probability of error in the case where the desired user (n ° 1) has sent the data “0”:

$$
\begin{aligned}
P_{e_{10}} &= prob\left(Z_i^{(1)} \geq S_1/b_i^{(1)} = 0\right) \\
&= prob\left(I_1 + A_1 \geq S_1/b_i^{(1)} = 0\right) \\
&= prob\left(A_1 = -1/b_i^{(1)} = 0\right) \cdot prob\left(I_1 \geq S_1 + 1/b_i^{(1)} = 0\right) \\
&+ prob\left(A_1 = 0/b_i^{(1)} = 0\right) \cdot prob\left(I_1 \geq S_1/b_i^{(1)} = 0\right)
\end{aligned} \tag{19}
$$

$$
\begin{aligned}
prob\left(I_1 \geq S_1 + 1/b_i^{(1)} = 0\right) &= f(S_1 + 1, N-2, N-2) \\
prob\left(I_1 \geq S_1/b_i^{(1)} = 0\right) &= f(S_1, N-2, N-2) \\
prob\left(A_1 = -1/b_i^{(1)} = 0\right) &= \frac{1}{2}\frac{W^2}{L} f(S_N, N-2, N-2) \\
prob\left(A_1 = 0/b_i^{(1)} = 0\right) &= 1 - \frac{1}{2}\frac{W^2}{L} f(S_N, N-2, N-2)
\end{aligned} \quad ;
$$

hence

$$P_{e_{10}} = \frac{1}{2}\frac{W^2}{L} f(S_N, N-2, N-2) \cdot f(S_1+1, N-2, N-2) + \left(1 - \frac{1}{2}\frac{W^2}{L} f(S_N, N-2, N-2)\right) \cdot f(S_1, N-2, N-2) \quad . \tag{20}$$

4.1.2 Second and $(N\text{–}1)^{th}$ cancelation stage

The decision variable for the desired user #1 is:

$$Z_i^{(1)} = Wb_i^{(1)} + I_2 + A_1 + A_2$$

where $I_2 = \sum_{k=2}^{N-2} \int_0^{T_b} b_i^{(k)} c_k(t) c_1(t) dt$, $A_1 = \left(b_i^{(N)} - \hat{b}_i^{(N)}\right) \int_0^{T_b} c_N(t) c_1(t) dt$, and $A_2 = \left(b_i^{(N-1)} - \hat{b}_i^{(N-1)}\right) \int_0^{T_b} c_{N-1}(t) c_1(t) dt$.

In this case, the term A_2 can take three values (−1.0 and +1):

$$A_1 = \begin{cases} -1 \text{ if an error is made on the data } b_i^{(N-1)} = 0 \left(\hat{b}_i^{(N-1)} = 1\right) \\ 0 \text{ if } \left(b_i^{(N-1)} = \hat{b}_i^{(N-1)} \text{ no error in detection}\right) \\ +1 \text{ if an error is made on the data } b_i^{(N-1)} = 1 \left(\hat{b}_i^{(N-1)} = 0\right) \end{cases} , \tag{21}$$

hence, $P_{e_2} = \frac{1}{2} P_{e21} + \frac{1}{2} P_{e20}$,
where $P_{e_{21}} = prob\left(\hat{b}_i^{(1)} = 0 / b_i^{(1)} = 1\right)$ and $P_{e_{20}} = prob\left(\hat{b}_i^{(1)} = 1 / b_i^{(1)} = 0\right)$

$$\begin{aligned} P_{e_{21}} &= prob\left(Z_i^{(1)} < S_1 / b_i^{(1)} = 1\right) \\ &= prob\left(W + I_2 + A_1 + A_2 < S_1 / b_i^{(1)} = 1\right) \\ &= prob\left(A_1 = 0 / b_i^{(1)} = 1\right) \cdot prob\left(W + I_2 + A_2 < S_1 / b_i^{(1)} = 1\right) \\ &\quad + prob\left(A_1 = -1 / b_i^{(1)} = 1\right) \cdot prob\left(W + I_2 + A_2 < S_1 + 1 / b_i^{(1)} = 1\right) \end{aligned} \tag{22}$$

and

$$\begin{aligned} P_{e_{20}} &= prob\left(Z_i^{(1)} \geq S_1 / b_i^{(1)} = 0\right) \\ &= prob\left(I_2 + A_1 + A_2 \geq S_1 / b_i^{(1)} = 0\right) \end{aligned} .$$

The calculation of the probability of error requires the determination of more and more terms. We have calculated the error probabilities by using the iterative method [8].

Figure 6 shows the theoretical and simulation bit error rate (BER) versus the threshold values of an OOC(73,4,1) for the conventional and the first cancelation stage of the Opt-SIC. We could see that the theoretical lines correlate with the simulation ones, and the Opt-SIC outperforms the traditional correlation receiver. From here on, we could validate the theoretical figures in order to show the performance of such receiver.

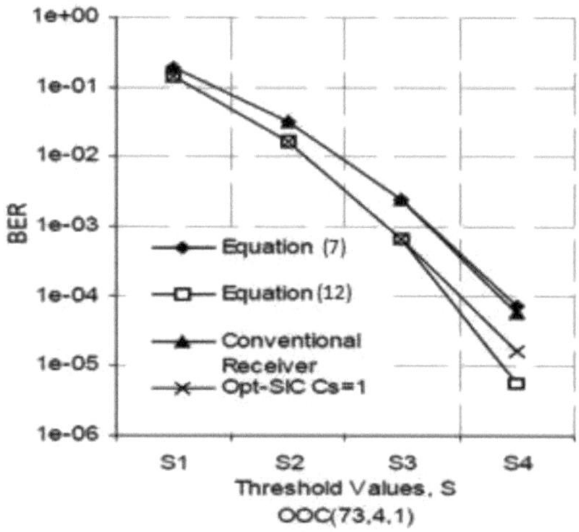

Figure 6.
Bit error rate (BER) versus the threshold values for a code length, F = 73 and for the conventional and the first cancelation stage.

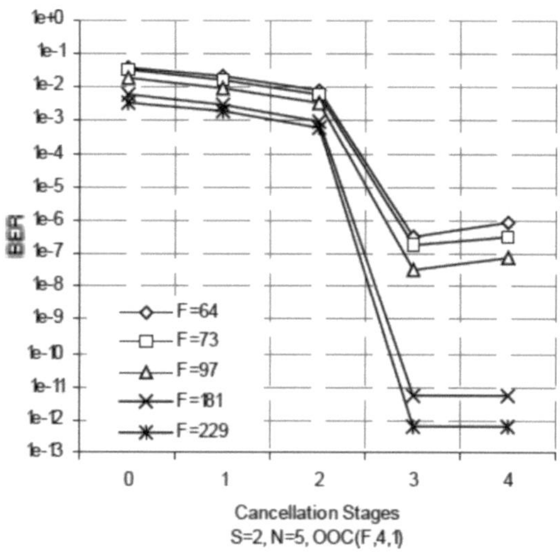

Figure 7.
The BER of the Opt-SIC system versus the cancelation stage for different code lengths and for threshold value equal to 2.

First, in order to evaluate the number of cancelation stages (Cs), we have studied the SIC performances with modifying the code weight. We can remark, in **Figure 7**, that when the number of cancelation stages increases, the BER first decreases slowly, and after a high decrease, it reaches a floor. Thus, it is not necessary to eliminate all N−1 stages in the SIC method to obtain a correct optical BER rate (10^{-9}). We can also just change the code length to obtain this achievement.

Moreover, in order to illustrate the benefit of the SIC in regard to conventional receivers, we have plotted on **Figure 8** the evolution of the BER as a function of the code weight W. We can observe that we obtain better performances with the SIC than with the CCR and the HL + CCR. Moreover, we can point out that the SIC is all the more performing when the BER decreases for the CCR.

In this section, an optical successive interference cancelation was examined, a technique based on the conventional O-CDMA receiver, in which the MAI is

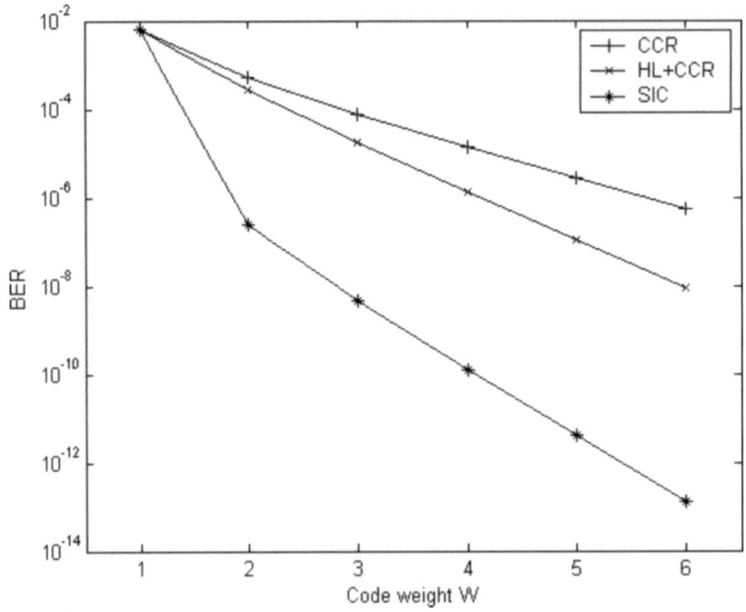

Figure 8.
BER as a function of the code weight W for F = 361 and N = 10, Cs = N−W + 1, SN = 4, and S1 = 3.

eliminated successively by a multistage cancelation system for the purpose of achieving a better performance. We have showed that the optical-SIC receiver manages to upgrade the performance of the conventional receiver by choosing the best threshold value, the cancelation stage, and the code length. It is important to note that, we could also improve the performance by using other parameters such as the code weight and the correlation values.

4.2 Parallel interference cancelation receiver (PIC)

4.2.1 Principles

The principle of a PIC receiver is based on the estimation of the interference due to all the nondesired users. Once the IAM is determined, it is discarded from the received signal before detecting the desired user. We start by detecting all the (N−1) nondesired users by a CCR receiver with a threshold level S_T $(0 < S_T \leq W)$. Each receiver generates the estimation $\hat{b}_i^{(j)}$ of the nondesired user # j data. This last one is spread by the corresponding code sequence. Then, the reconstructed interference is removed from the received signal. Finally, the data of the desired user are detected with a CCR with a threshold level S_F $(0 < S_F \leq W)$ (**Figure 9**) [9].

4.2.2 PIC receiver

The first stage of parallel interference cancelation receiver is a parallel structure whose receivers are all CCR. The signal applied to the second stage is expressed as:

$$\begin{aligned} s(t) &= r(t) - \sum_{j=2}^{N} \hat{b}_i^{(j)} c_j(t) \\ &= \hat{b}_i^{(1)} c_1(t) + \sum_{j=2}^{N} \left(b_i^{(j)} - \hat{b}_i^{(j)} \right) c_j(t) \end{aligned} \tag{23}$$

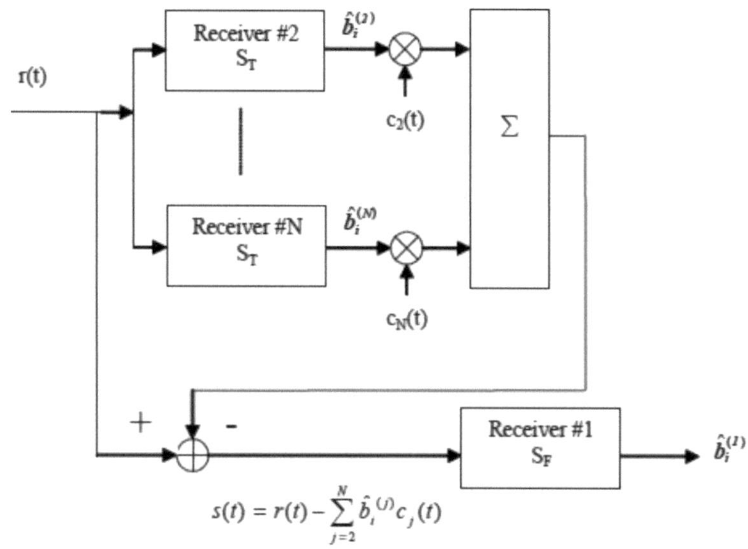

Figure 9.
Parallel interference cancelation structure.

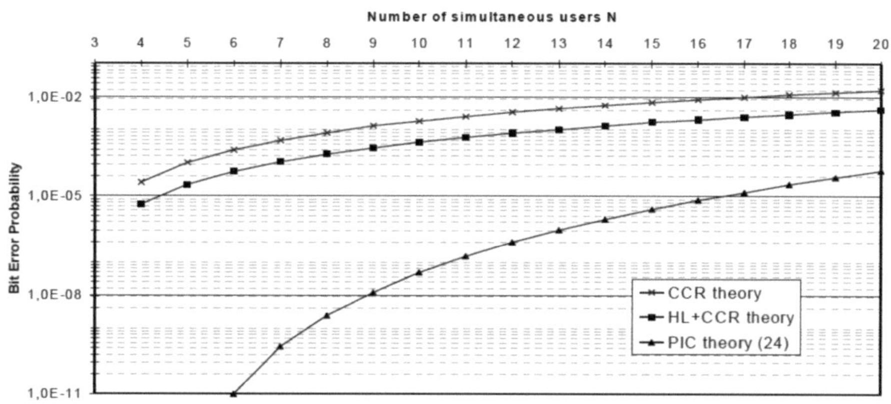

Figure 10.
BER as a function of the number of simultaneous users N:F = 121 and W = 3.

where $\left(b_i^{(j)} - \hat{b}_i^{(j)}\right)$ can take two values: "0" or "−1". So, the second term in Eq. (23) generates negative interference on user #1. Thus, one of the structure originalities is that errors can occur only when the desired user's sent datum is 1, contrary to CCR.

We have established in [8] that the theoretical expression of the PIC error probability is:

$$P_{ePIC} = (1/2)^N . \sum_{N_1=S_T-1}^{N-1} \sum_{N_2=W+1-S_F}^{N-1-N_1} \binom{N-1}{N_1}\binom{N-1-N_1}{N_2}(P_{IPIC})(1-P_{IPIC})\left(\frac{W^2}{F}\right)^{n_1}\left(1-\frac{W^2}{F}\right)^{N_1-n_1}$$

$$\text{with } P_{IPIC} = \frac{W^2}{F}\sum_{n_1=S_T-1}^{N_1}\binom{N_1}{n_1}\left(\frac{W^2}{F}\right)^{n_1}\left(1-\frac{W^2}{F}\right)^{N_1-n_1}. \qquad (24)$$

A theoretical comparison of the performances of a CCR, a CCR with an optical hard limiter and a PIC receiver was made according to the number of simultaneous

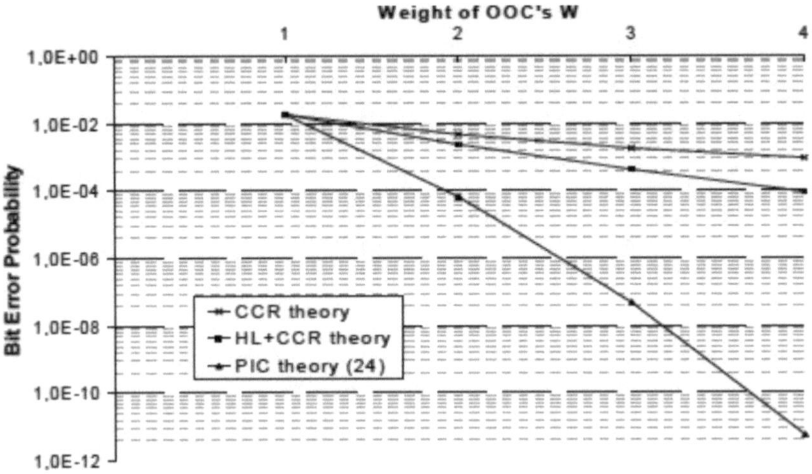

Figure 11.
Bit error probability versus the weight of OOCs W:F = 121 and N = 10.

users for an OOC (F = 121, W = 4) code. The bit error rate versus the number of simultaneous users is presented in **Figure 10**. We can observe that the PIC receiver has better BER than the conventional receiver. So, the PIC receiver can support more users than the conventional ones. For example, for BER equal to 10^{-5}, the PIC can support N = 16 users, while the conventional receiver with hard limiter can support only N = 4 users.

The BER versus the weight W for DS-CDMA systems for a CCR with and without a hard limiter, and a PIC receiver is plotted on **Figure 11**. F = 121 and N = 10 have been considered as the OOC's code parameters. According to the simulation results, we note that the PIC receiver has a better BER than the traditional ones. Furthermore, at a bit error probability of 10^{-4}, for example, the PIC needs a weight of W = 2, while the conventional receiver with hard limiter needs W = 4. This configuration of receiver shows so many advantages. First of all, the required power is reduced. In a second time, as the weight is reduced, the number of potential users increases. On the one hand, with W = 4, we have 10 available code sequences, so with N = 10 users, all the sequences are used. On the other hand, W = 2 corresponds to 60 available code sequences; thus, there are 50 unused sequences. Consequently, we can build a system where the sequence codes are definitively assigned to the users, but only 10 of them are allowed to simultaneously communicate.

In order to reduce the effect of the only limitation of an ideal DS-OCDMA system (without taking into account the impact of optoelectronic components), which is multiple access interference (MAI), we have studied the performance of an optical parallel interference cancelation receiver.

This study is based on the analytic expression of the error probability and the system simulation, a comparison with the conventional receiver has been made in each case. The results found prove that the interference cancelation receiver outperforms the ones of conventional receiver.

The PIC receiver could be a suitable receiver in the case of highly loaded networks.

5. Fiber chromatic dispersion effects on OCDMA

In the previous paragraphs, we have showed that the performances of a DS-OCDMA system are degraded by the MAI. However, the MAI is not the only restriction in the OCDMA optical link networks. In reality, in an optical transmission system, several physical phenomena can degrade the performances of the system [10–12]. The main limitation is due to the chromatic dispersion of the fiber. Indeed, chromatic dispersion leads to optical pulse broadening. This broadening results in overlapping between the pulses, which create the interference inter different chip transmitted over optical fiber, which can affect the system performances. At the output of the optical fiber, it is possible to express the electric field of the jth chip (code) of the lth bit (data) as a function of the in-phase component and quadrature as follows:

$$E^1_{l,j} = \left(y^{l,j}_{p1} + jy^{l,j}_{q1}\right) + \left(\sum_{i=2}^{N} y^{l,j}_{pi} + j\sum_{i=2}^{N} y^{l,j}_{qi}\right) + \left(\sum_{h=-\left\lceil\frac{T-T_c}{2T_c}\right\rceil}^{\left\lceil\frac{T-T_c}{2T_c}\right\rceil} y_{Dp_{j-h}} + j\sum_{h=-\left\lceil\frac{T-T_c}{2T_c}\right\rceil}^{\left\lceil\frac{T-T_c}{2T_c}\right\rceil} y_{Dq_{j-h}}\right) \quad (25)$$

where (y_{pi}, y_{qi}) are the components, respectively, in phase and quadrature of the electric field of the user i^{th} and (y_{Dp}, y_{Dq}) those due to the chromatic dispersion. The first term corresponds to the data of the desired user, the second is the corresponding term to the MAI, and the last is due to the chromatic dispersion. Indeed, the superposition of the MAI term and the term due to the dispersion produce a sufficient power to degrade the system performances.

The propagation distance is usually short in the access networks. For such networks, we deploy the G652 single-mode fiber optics. Therefore, intramodal dispersion can be neglected. Asymmetries and stress distribution in fiber core, which leads to birefringence, cause the polarization mode dispersion (PMD). It affects only long-haul communication systems. Nonlinear effects (Kerr and Raman effects) in optical fiber can degrade the performances of the system, but mainly for long-distance communication. Such nonlinearities are dependent on the signal intensity, which are not significant at the low power. Therefore, for a short access optical link, chromatic dispersion effect is an important factor, which needs to be addressed. For high data rates, the OCDMA technique requires the generation of ultrashort pulses. Indeed, for a given data rate D, the chip rate Dc is expressed as: Dc = 1/Tc = F.D. These pulses are sensitive to chromatic dispersion.

To study the impact of chromatic dispersion on the performances of the system, several simulations have been made in the case when there is no dispersion compensation component deployed. The study has been done according to different network parameters and those of the code.

At first, we analyze the performances as a function of the fiber length for a fixed pulse duration Tc = 1/(F * D) = 2.51 10^{-11} s and the number of active users N = 5.

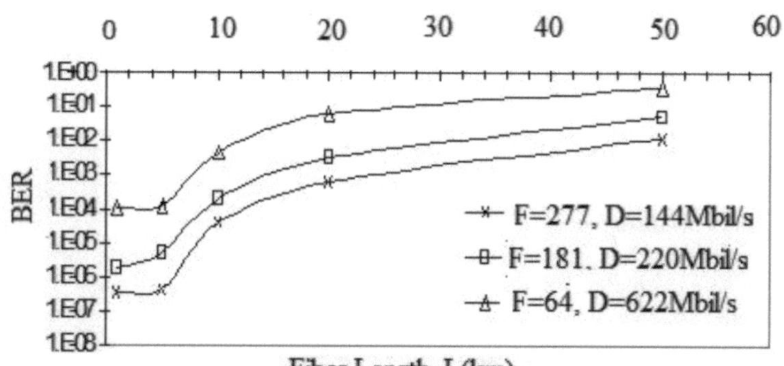

Figure 12.
Variation of BER as a function of fiber length, where Tc (F,D) = 2.51 10^{-1} s and N = 5.

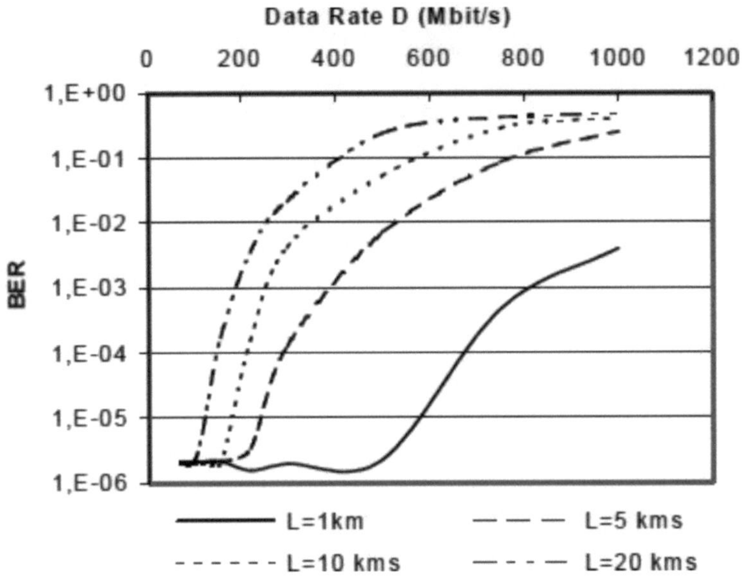

Figure 13.
BER versus data rate D for an OCDMA link with OOC (F = 181, W = 4) and N = 5 active users.

Figure 12 illustrates the variation of the BER as a function of the fiber length for three codes with various Fs and Ds. It can be noted that the dispersion effect increases as the fiber length increases. However, for this particular chip duration, the dispersion has no impact on the BER for optical fibers shorter than 5 km. On the other hand, when the fiber's length is greater than 5 km, system performance is deteriorated.

In order to complete our study, the BER versus data rate D, for a code length F = 181, has been simulated (**Figure 13**). For example, on the one hand, we can observe that the performances of an OOC (F = 181, W = 4) are not affected by the fiber dispersion up to D = 600 Mbits/s for a 1-km-long optical link. On the other hand, for a 20-km-long optical link, the performances are degraded from a data rate D = 100 Mbits/s.

The curves indicate that for making the effect of dispersion negligible, without using in the OCDMA link, a dispersion compensated component, we should have a trade-off between OOC code length F and data rate D.

The parametric study has shown that the OCDMA link performances in the access network context are significantly overestimated when the fiber chromatic dispersion is neglected.

6. Conclusion

In this chapter, we studied the DS-OCDMA multiple access technique envisaged for optical communications, in particular in PON access networks. To maintain high rates, the code spreading length should be as low as possible. In this case and for an incoherent system, it has been shown that the IAM multiple access interference, linked to the use of quasi-orthogonal unipolar codes, is very important and does not make it possible to maintain the quality of the link. It is, therefore, necessary to reduce the MAI in order to be closer to the specifications. For this purpose, two structures were studied: the serial interference cancelation (SIC) and the parallel interference cancelation (PIC). We first developed the approximate theoretical expression of the SIC error probability for unipolar codes whose intercorrelation is equal to 1. We have verified that SIC improves performance with respect to the conventional correlation receiver (CCR). Due to the dependency of a stage on previous ones, the exact theoretical analysis is very difficult to carry out.

Then, the efficiency of the PIC structure for unipolar codes whose intercorrelation is '1' has been investigated.

We have studied theoretically and by simulation the performance of a system using OOC codes, and we have shown that the PIC can significantly improve the performance.

Finally, since IAM is not the only limitation of performance, we have studied the impact of chromatic dispersion. It is demonstrated that chromatic dispersion has a significant negative effect on system performance, which cannot be neglected for systems with a short fiber length and a high data rate. It is reported that system performance can be significantly overestimated if chromatic dispersion is ignored.

Appendices and nomenclature

ADSL	asymmetric digital subscriber line
CCR	conventionnal correlation receiver
CDMA	code-division multiple access
DS-CDMA	direct sequence-CDMA
FDMA	frequency-division multiple access
FFH-CDMA	fast FH-CDMA
FTTB	fiber to the building
FTTC	fiber to the curb
FTTH	fiber to the home
HL	hard limiter
MAI	multiple access interference
OCDMA	optical CDMA
OOC	optical orthogonal code
PIC	parallel interference cancelation receiver
PMD	polarization mode dispersion
PON	passive optical network
SIC	successive interference cancelation receiver
TDMA	time-division multiple access
WDMA	wavelength-division multiple access

Author details

Younes Zouine and Zhour Madini*
Department of Electrical and Telecommunication, ISET Laboratory ENSA, Kenitra, Morocco

*Address all correspondence to: zmadini@gmail.com

References

[1] Salehi JA. Code division multiple-access techniques in optical fiber networks–Part I and part II. In: IEEE Transactions on Communications., vol 37, n°8, pp. 824-842, 1989. DOI: 10.1109/26.31181

[2] Stok V, Sargent E. Lighting the local area: Optical code-division multiple access and quality of service provisioning. IEEE Network. Nov/Dec 2000;**14**(6):42-46. DOI: 10.1109/65.885669

[3] S. Zahedi, J. A. Salehi Performance analysis for various fiber-optic CDMA receivers structures. Journal of Lightwave Technology, vol 18, n°12, pp. 1718-1727, 2000. DOI: 10.1109/GLOCOM.2000.891329

[4] Chen JJ, Yang GC. CDMA fiber-optic systems with optical hard-limiter. Journal of Lightwave Technology,vol 19, n°7, pp 950–958, july 2001. DOI: 10.1109/50.933289

[5] Tang JTK, Letaief KB. Optical CDMA communications systems with multiuser and blind detection. IEEE Transactions on Communications,vol 47 ,n°8, pp 1211–1217, august 1999. DOI: 10.1109/26.780457

[6] Fantacci R, Tani A, Vannuccini G. Performance evaluation of an interference cancellation receiver for noncoherent optical CDMA systems. GLOBECOM. 2002;**3**:2823-2827. DOI: 10.1109/GLOCOM.2002.1189144

[7] Verdù S. Multiuser Detection. New York, USA: Cambridge University Press; 1998

[8] Zouine Y, Saad NM, Goursaud C, Vergonjanne A, Aupetit-Berthelemot C, Cances JP, et al. The influence of the optical successive interference cancellation in the optical CDMA network, ISSLS 2004: XVth International Symposium on Services and Local Access. 2004 March 21-24; Edinburgh, UK

[9] Goursaud C, Saad NM, Zouine Y, Julien-Vergonjanne A, Aupetit-Berthelemot C, Cances JP, et al. Parallel multiple access interference cancellation in optical DS-CDMA systems. Annals of Telecommunications. 2004;**9**(10):1053-1068

[10] Zouine Y, Dayoub I, Haxha S, Rouvaen JM. Analyses of constraints on high speed optical code division multiplexing access (OCDMA) link parameters due to fiber optic chromatic dispersion. Optics Communication. 2008;**281**:1030-1036. DOI: 10.1016/j.optcom.2007.10.079

[11] Zouine Y, Madini Z, Aloui KS, Foshi J. Performance analysis of parallel interference cancellation receiver applied to hybrid OCDMA/WDM system for high-speed optical access. In: 2015 Asia-Pacific Microwave Conference (APMC). December 2015. DOI: 10.1109/APMC.2015.7413184

[12] Zouine Y, Madini Z, Aloui KS, Foshi J. DS-OCDMA/WDM receivers based on parallel interference cancellation for gigabit passive optical network. International Journal of Microwave and Optical Technology. 2017;**12**(3):213-220. ISSN: 1553-0396

Chapter 6

Optically Multiplexed Systems: Wavelength Division Multiplexing

Meena Dasan, Fredy Francis, Kundil T. Sarath, Elambilayi Dipin and Talabattula Srinivas

Abstract

Optical multiplexing is the art of combining multiple optical signals into one to make full use of the immense bandwidth potential of an optical channel. It can perform additional roles like providing redundancy, supporting advanced topologies, reducing hardware and cost, etc. The idea is to divide the huge bandwidth of optical fiber into individual channels of lower bandwidth, so that multiple access with lower-speed electronics is achieved. This chapter focuses on one of the most common and important optical multiplexing techniques, wavelength division multiplexing (WDM). The chapter begins with a quick historical account of the origin of optical communication and its exponential growth following the invention of erbium-doped fiber amplifier (EDFA) leading to the widespread adoption of WDM. Alternate multiplexing schemes are also briefly discussed, including time-division multiplexing (TDM), space-division multiplexing (SDM), etc. A typical WDM link and its components are then discussed with special focus on WDM Mux/demultiplexer (DeMux). Further, certain challenges in this field are addressed along with some potential solutions. The chapter concludes by highlighting some features and limitations of optically multiplexed WDM systems.

Keywords: WDM, Mux, DeMux, EDFA transients, optical components

1. Introduction

Since its advent in the mid-1960s, optical technologies and components have been changing the landscape of communication as such. The constant push for higher data rates ensured that optical components matured fast, enabling the terabits of data rates we are enjoying today. It all started with extremely lossy optical fibers coupled with a broadband source, which could transmit only a few mbps over a few meters. The scenario drastically changed over half a century, reaching data rates of even terabit/sec possible over a single fiber. Now we have the luxury of external cavity lasers (ECL) which can easily give linewidths below 1 MHz, Mach-Zehnder modulators (MZM) which can easily operate at 40 Gbps and above, low attenuation dispersion managed fibers, dispersion compensation fibers, low-noise high-gain optical amplification systems, very high-speed detectors, and extremely fast digital signal processing (DSP) capabilities which make many compensation hardware redundant.

A commonly overlooked component of this lot is the multiplexer, without which the entire system is only fast enough as any of its electronic counterpart. Multiplexer is the one which helps in combining (and splitting) the data from different sources so that the tremendous bandwidth of the system could be utilized. As with other components, Mux/DeMux came a long way into maturing and providing the kind of precise work it does today. So, this chapter is dedicated for giving the readers a quick understanding of the different types and techniques for the implementation of wavelength division multiplexing (WDM) schemes.

Multiplexing is the process of combining multiple signals into a shared channel used for tapping the full potential of the optical links. It facilitates networking with advanced topologies supported with redundancy features. Historically, multiplexing had been used to share the limited bandwidth of the medium between different transmitters, but with optical systems it is more about making full use of the huge available bandwidth. This is where wavelength division multiplexing comes in where different channels are multiplexed into a single fiber. It divides the huge bandwidth of optical fiber into various logical channels of lower bandwidth that can be filled with electronically achievable data rates. The advent of coherent optical links and advanced multiplexing techniques used in wireless communication raised the achievable bandwidth limit of fiber links. But the proposed chapter focuses on one of the most common and important multiplexing techniques, wavelength division multiplexing.

The advent of erbium-doped fiber amplifiers (EDFAs) in the late 1980s proved to be a great impetus for multichannel system implementation. EDFA, with its capability to amplify the C-band, proved to be a vital component that could facilitate WDM in long-haul links. As the different WDM channels could traverse the fiber without cross talk, and EDFA can amplify these signals simultaneously, it increased the transmission rates exponentially. This ushered in the need of multiplexers, specifically wavelength division multiplexers. A few popular optical multiplexing techniques are discussed later in this chapter. Also, it should be noted that being bidirectional, most of these devices in these schemes can be used as Mux or DeMux.

1.1 Other optical multiplexing techniques

With the optical components maturing and becoming very reliable and accurate, almost all multiplexing techniques possible in the wireless communications are viable in optics now. Though a few are truly developed, the others are expected to mature in the near future. A few of these techniques are discussed here.

1.1.1 Time-division multiplexing

Probably the most used scheme in electrical and wireless systems, optical time-division multiplexing (OTDM) does not have that much widespread use, probably because of the large bandwidth already available with optical fibers and the widespread use of WDM but is still used in different applications. Pulse interleaving is used to allow different optical data streams to use the full capacity of the fiber, albeit for a limited time. Different optical data pulses thus share the same optical channel with its full capacity for a limited time and passes on to the next pulse and so on so forth. Thus, the overall data rate in the fiber is improved despite the fact that the individual data streams are still operating at the same speed. But the optical pulses have to be compressed, to fit in the time slot per bit. A train of ultrashort pulses can be helpful in this regard, along with an optical delay line for combining all the pulses. So, for a good OTDM link, we need pulses with high extinction, short duration, and low timing jitter.

1.1.2 Space-division multiplexing

The basic idea is to have different channels separated spatially. This spatial separation can be brought out in different ways; the simplest is with the use of multiple fibers. But this requires most of the channel infrastructure to be duplicated for each fiber, hence not most economical. But still there could be some cost improvements as one may choose to use a high-power laser along with splitters to pump different optical amplifiers corresponding to each fiber. Or multiple fibers can be integrated to form one fiber cable or use a fiber ribbon. Another possible way of cost reduction is by using a vertical-cavity surface-emitting laser (VCSEL) array at the source along with the integration of different receivers into the same chip. Also, some electronics can also be shared between different channels too.

A better implementation of SDM will be using multicore fibers. These are single optical fibers having multiple cores, each core carrying one communication channel. The cores are usually separated enough that there is no coupling and cross talk. But this limits the number of cores, as increasing the cladding diameter tends to make the fiber more brittle. Additionally, this makes fiber splicing more difficult.

A work around of this is to keep the core together allowing the different modes to couple to form a supermode (coupled SDM fiber) and then later electronically process the detected data using multiple input/multiple output (MIMO) techniques in detection process. MIMO techniques have already matured in wireless communication. Photonic crystal fibers are a very good candidate for these kinds of coupled SDM fibers. But there are a lot of challenges that are to be dealt with before this becomes commercial; this includes difficulties in splicing multicore fibers, maintaining the correct rotation orientation, developing spatial multiplexers and fiber amplifiers, providing sufficient gain uniformity over different modes, etc.

Another potential method for the MIMO setup is the use of few-mode fibers, using a technique called mode division multiplexing (MDM). This has some advantages like easier fiber connections and splices. But as the cross talk with different fiber modes are high and there is a significant difference in the group velocity of the different modes, MIMO receivers are much more complex than the above techniques [1].

1.1.3 Polarization-division multiplexing

Generally used together with phase modulation or QAM, the idea is to modulate different information on orthogonal polarizations of the same frequency effectively doubling the data rate. PDM signal can be transmitted over normal WDM infrastructure expanding its capacity. But here the challenges are the drifts in the polarization state of the fiber-optic system over time. Also polarization mode dispersion, polarization-dependent losses, and cross-polarization modulation create additional challenges. But the DSP techniques at the receiver are becoming faster and more efficient in compensating these impediments.

1.1.4 Orbital angular momentum multiplexing

The orbital angular momentum (OAM) of the light waves is used to carry orthogonal information. OAM can in theory have huge number of parallel channels, bound only by practical limitations. OAM states of light is not supported on conventional single-mode fiber, instead few-mode or multimode fibers are to be used. Additionally, conventional fiber suffers from mode coupling which changes the orbital angular momentum when the fiber is bent or stressed causing mode instability. But coherent detection along with signal processing techniques can be used to correct mode mixing in fiber. These coherent systems are normally complex in nature.

2. WDM link and components

A WDM link typically used in communication system is as shown in the block schematic (**Figure 1**). There are multiple transmitters each working on its own dedicated wavelength. These individual data streams on independent wavelengths are all multiplexed together using a WDM Mux and transmitted through an optical fiber. The transmitted power is kept low enough in the link, so as not to trigger any nonlinearities in the fiber. In the absence of nonlinearities, the wavelengths do not talk to each other or induce cross talks. The C-band can be divided coarsely every 20 nm (called coarse WDM) if the link is low cost or have to support only limited number of parallel light paths. Dense WDM uses a much more tighter wavelength division scheme and needs costlier components and better lasers with lower linewidths and external modulators. It supports a greater number of channels resulting in a much higher throughput, e.g., 40 channels within 1530–1565 nm, C-band with 100 nm spacing, or 80 channels at 50 GHz spacing or even 12.5 GHz—all at the cost of significant overhead costs.

These multiplexed data will become weaker over the distance, and that's where the EDFA comes in. It's EDFA's broad bandwidth, which helps in amplifying all the channels with nearly the same gain, that paved the path for the bandwidth explosion within the fiber. After the amplification the wavelengths are demultiplexed at the DeMux and forwarded to the corresponding receivers which completes the link.

Typical WDM link consists of components like transmitters, add/drop multiplexers, and necessary detectors for the communication link. Based on the requirements, it also includes preamplifier, line amplifiers, and post amplifiers in the link. Specific WDM components like WDM/demultiplexer (Mux/DeMux) matured fast to make the systems relatively common and affordable. It also made possible to have fiber links to their maximum possible bandwidth capability.

2.1 Optical source

Laser is the most widely used optical source, owing to its inherent advantages like single-frequency operation, coherence, high intensity, and directionality [2]. Laser is essentially an oscillator, having a gain medium and feedback mechanism. The gain medium is kept at an excited state by external pumping mechanisms (optical or electrical). This produces population inversion which is a necessary condition for lasing. Initial photons can be introduced into the medium by spontaneous emission, and those modes that are supported by cavity and gain spectrum are lasered out. The basic assembly of a laser is as shown in **Figure 2**.

The development of semiconductor lasers operating at room temperature (1970) provided a compact, highly efficient, and reliable optical source, which is put to great use by the communication industry [3]. Semiconductor laser needs

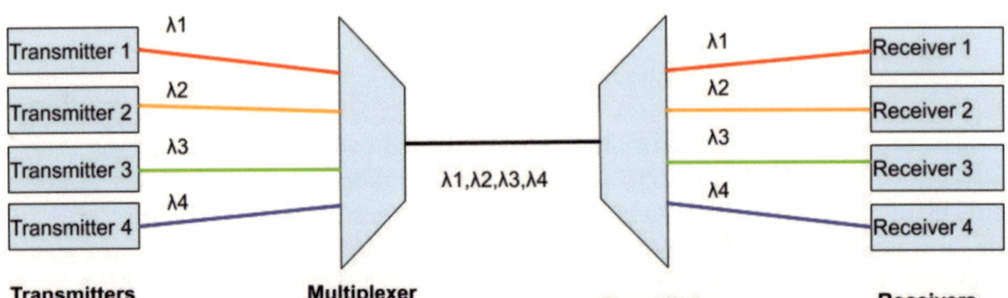

Figure 1.
Block schematic of a basic WDM link.

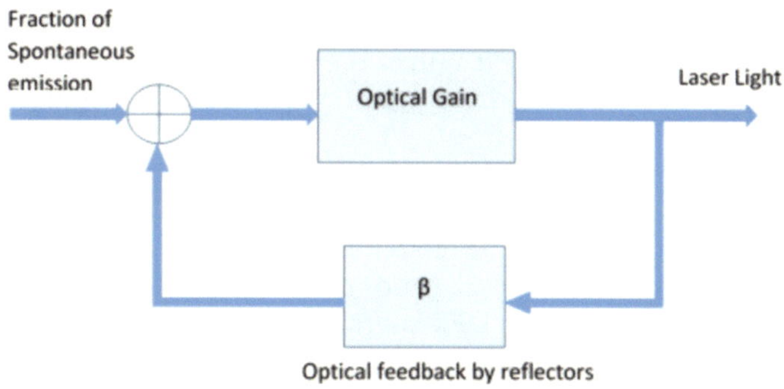

Figure 2.
Laser assembly block schematic.

an optical feedback mechanism for converting it from an amplifier to oscillator. Oscillations begin when the loop gain exceeds unity. Gain is obtained due to stimulated emission in optical gain medium, and the cavity formed by cleaved laser facets provides the required feedback for sustained oscillations. This configuration forms a Fabry-Perot (FP) cavity. FP laser can lase at multiple longitudinal modes that are spaced apart according to $\Delta \nu_L = c/2L$ where n is the refractive index (RI) and L is the cavity length. If the spacing between neighboring modes are small enough, then the laser cavity can provide almost the same gain for each of those longitudinal modes causing them to coexist. Such longitudinal modes travel with a different velocity inside the optical fiber causing group velocity dispersion and hence limits the maximum data rate through the optical fiber. So, lasers operating within a single longitudinal mode are preferred for many applications. To overcome this, distributed feedback (DFB) lasers, which achieve single longitudinal mode operation by distributing the reflection throughout cavity length, are introduced [3].

DFB laser developed during the 1980s is the most commonly used single-mode laser. The idea is to introduce a wavelength selective element within the laser cavity. This is achieved by introducing an etched diffraction grating within the laser waveguide structure. This can be done in two ways. If the grating layer extends through the whole of the active layer, it is called DFB, and if the gain region is in a separate planar section, the device is known as distributed Bragg reflector (DBR). **Figure 3** shows the structure of a DFB laser.

In DBR lasers, the fiber Bragg gratings are used like mirrors in FP cavity with the difference that these gratings reflect only one longitudinal mode. The active layer provides cumulative gain only for the feedback wavelength, hence resulting

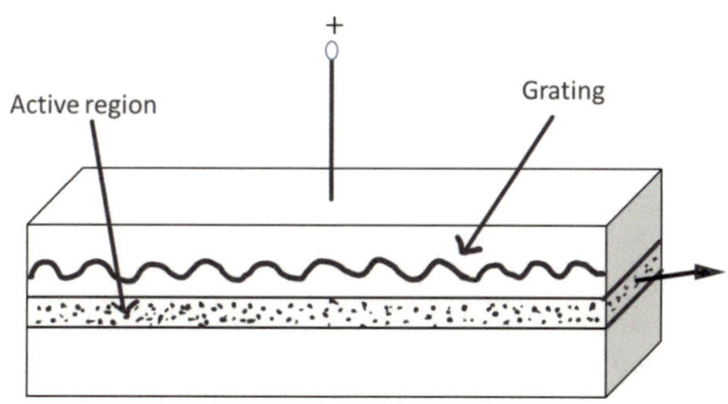

Figure 3.
DFB laser structure [3].

in single-wavelength operation of laser. A Bragg grating is realized by periodically varying refractive index along the length of the optical transmission. Condition for reflection of Bragg wavelength λ_B from a grating with period Λ is given by

$$\Lambda = m\ (\lambda_B/2n) \tag{1}$$

where m is an integer and n is the refractive index. Due to frequency selective nature of the feedback mechanism, the output of the laser becomes highly monochromatic. Later improvements in DFB lasers include phase-shifted DFB laser [4] and gain-coupled DFB lasers [5].

In a semiconductor laser under forward bias, population inversion occurs, and optical gain is realized only when the injected current into the active region is greater than a minimum value known as the transparency value [3]. The input signal propagating inside the gain medium is amplified by a factor of e^{gz}, where g is the gain coefficient and z is the length within the cavity. A certain portion of the generated photons is lost due to cavity losses and needs to be replenished. The optical gain must be high enough to compensate for this loss, else the photon population does not build up. This puts a minimum value of gain with which the laser should be operated to achieve lasing. This is the laser threshold, which is achieved only if the laser works above a threshold pumping level which corresponds to the threshold current.

2.2 Modulators

The process of imposing data on the light stream is called modulation. At bit rates more than 10 Gb/sec, chirp effect induced by direct modulators is predominant and puts a limit on modulation bandwidth. Chirp is a phenomenon wherein the carrier frequency of transmitted pulse varies with time, causing broadening of the transmitted spectrum. Laser output acts as a carrier signal over which the input signal gets modulated with the help of modulators. These modulators are classified into electro-absorption (EA) and electro-optic modulators. The performance of an external modulator is measured based on extinction ratio and the modulation bandwidth.

In direct modulation the laser output intensity is controlled by directly modulating the injection current of laser diode in accordance with the input signal. But this can lead to chirping effect at higher frequencies. So, it is preferred to keep the laser source as itself and modulate the light output from it by keeping an external modulator in front. **Figure 4** shows the schematics of direct and external modulation schemes.

2.2.1 Direct modulation

Direct modulation of laser drive current is simpler, cost effective, and gives satisfactory performance for lower-frequency-modulating signals. But as the drive current to laser is varied directly, turn on delay and oscillation can result out of the fast-changing pumping current causing frequency chirping and linewidth broadening [6]. ON and OFF operations of laser cause the gain to change rapidly in the lasing medium. The change in gain causes a change in carrier concentration which in turn changes the refractive index, and this periodic change in the refractive index results in frequency chirping (the spectrum changes with time). When a chirped pulse propagates through a dispersive medium like optical fiber, the spread in frequency causes certain portion of the wave to travel faster/slower with respect to other portions leading to intersymbol interference (ISI).

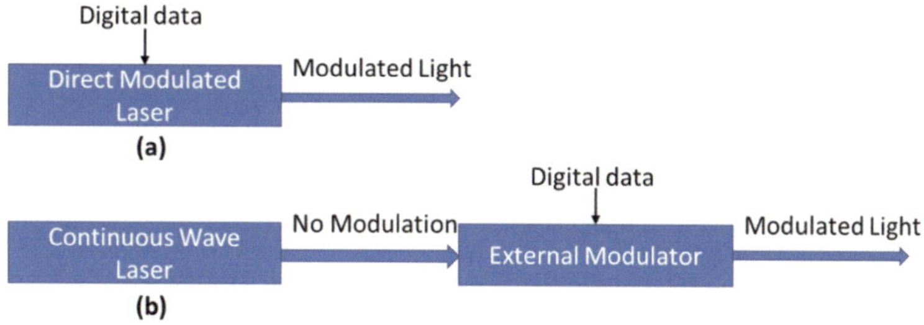

Figure 4.
Modulation schemes (a) direct modulation and (b) external modulator.

2.2.2 External modulation

A direct modulation is not suited for very high data rate transmission due to reasons such as chirp in DFB laser and mode partition noise in FP laser. So external modulation schemes are used for high-frequency modulation as it does not affect the laser characteristics and is implemented as an additional component in front of a CW laser. But this leads to an additional insertion loss for external modulators. Large modulation bandwidth and depth, small insertion loss, lower electrical drive power, etc. far outweigh its cons for bit rates above some 10 Gb/s. Some desirable characteristics of external modulators are polarization independence, good linearity (between drive current and modulated output), lower cost, and smaller size.

External modulators make use of techniques like electro-optic effect, acousto-optic effect (AOF), and electro-absorption effect to modulate the information signal over the incident CW optical beam. Acousto-optic modulators are slower and hence are commonly not used for communication purposes. *Electro-absorption* modulators are usually II–V materials, which alter its absorption coefficients according to an external voltage to obtain intensity modulation directly. These modulators are generally capable of attaining an extinction ratio of 15 dB or more at bit rates up to 40 Gb/s [7]. They have further advantages of being efficient and compact in size and can also be easily integrated into the same chip along with laser, as these modulators are also made of the same material.

Another class of external modulators are the *electro-optic modulators*, which works by altering its optical properties (mostly the refractive index (RI)) with the application of electrical field. This refractive index change may be due to Pockels effect, where the RI changes linearly with the applied electric field, or due to Kerr effect wherein the RI change is proportional to square of the electric field amplitude. When an optical signal passed through the altered RI region, it induces a phase change (or polarization rotation) in the signal, which can be converted to amplitude modulation by using Mach-Zehnder interferometer (MZI) configuration. LiNbO3 is the most widely used medium for this purpose, as it is optically birefringent and hence can be externally controlled by an applied electric field. A Mach-Zehnder-type external modulator uses phase modulation along with an integrated Mach-Zehnder interferometer to achieve intensity modulation. These modulators are capable of modulating up to 60 GHz [7]. The modulating frequency can be further extended up to 100 GHz or higher with traveling wave electrode configuration [8].

2.3 WDM multiplexer and demultiplexer

The inherent immense bandwidth of optical fiber systems can be tapped by the use of multiplexing techniques, which facilitates the electronics to work in much lower rate than the optical transmission rate. It is known that transmitting data over a single fiber with higher rates is more economical than carrying lower data rates over several fibers [9]. This makes multiplexing a must, so that huge transmission capacity of the optical fibers can be supported using moderate electronic component rates.

Different varieties of Mux/DeMux are available. Mostly these are reciprocal devices, hence can be used as both Mux and DeMux. They can be classified under two broad categories, diffraction-based and interference-based. Diffraction-based devices rely on angular dispersive element like diffraction gratings to decompose the incident light into its spectral components.

Interference-based DeMux are based on optical filter and directional couplers. Filter-based Mux uses optical interference for wavelength selectivity. MZI-based filters are the most used. One arm of MZI can be made longer to induce phase difference with respect to the other arm. This phase difference is frequency dependent. The path length is adjusted such that power from two different input ports adds up at only one output port (Mux operation).

The idea is to spatially separate the different wavelength that were traversing together in the optical fiber. Each of these wavelengths can be collected in those points into individual optical fibers. Optical Mux/DeMux can be broadly classified into passive and active. Popular passive Mux/demultiplexers are based on Prims, diffraction gratings, and spectral filters. Active Mux/DeMux is usually implemented as some passive components along with tunable detector, each tuned to separate wavelength. We will see each of them in a bit more details now.

2.3.1 Prism-based devices

These devices work based on the principle of dispersion, where different wavelengths see a different refractive index in the medium. This difference in refractive index results in some wavelengths to bend more (or less) than others which helps in separating them out. As can be seen in **Figure 5**, the incoming wavelengths are collimated and incident on the prism. Each wavelength seems a slightly different refractive index and bends differently according to Snell's law. At the output another lens focuses the different wavelengths to different output fibers.

2.3.2 Superprism-based devices

Superprisms employ photonic bandgap that make certain wavelength forbidden within the structure. This is achieved using special structures called photonic crystal. A photonic crystal is a periodic dielectric structure fabricated usually on Si using nanofabrication. This three-dimensional periodicity in refractive index causes periodic distribution in bands and gaps, and these can be tuned by varying the periodicity so as to make certain wavelength to propagate or not. It can act both

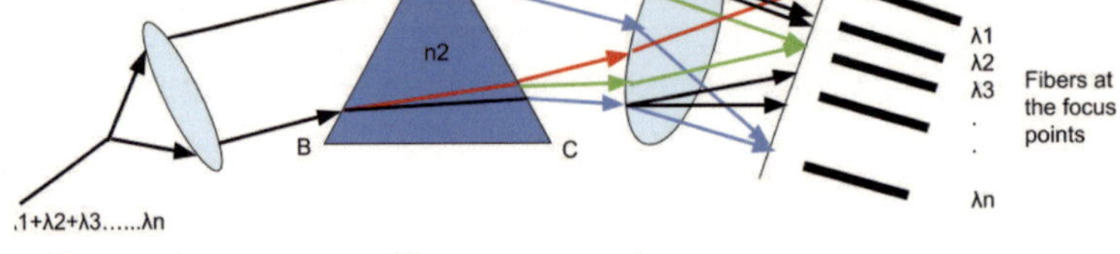

Figure 5.
Prism-based DeMuX configuration [10].

as energy bandgap filters as described above and as highly dispersive media. This high-dispersion property can be used to make prism called superprisms as they have almost 500 times more dispersion than normal prisms.

2.3.3 Diffraction grating-based devices

Diffraction elements as the name suggests use diffraction of light to separate different wavelengths. When a polychromatic light wave is incident on a diffraction grating, each wavelength is diffracted at a different angle from where they can be collected to achieve demuxing (**Figure 6**).

2.3.4 Arrayed waveguide grating (AWG)-based devices

AWG works on the principle of interference on a specially designed structure as shown in **Figure 7**. It has two free space propagation regions (S1 and S2), an array of waveguides (Wn) in the middle and fibers for input and output. A WDM signal incident on S1 through F traverses the free space and enters the arrayed waveguide region. The length of each waveguide in the arrayed waveguide section is varied such that it introduces a wavelength-dependent phase delay in S2. This phase delay causes the interference points of each wavelength to be spatially separated, where a fiber is connected to collect each wavelength, hence attaining DeMux.

AWG has some interesting features as follows which makes it very attractive.

- AWG has a flat spectral response.

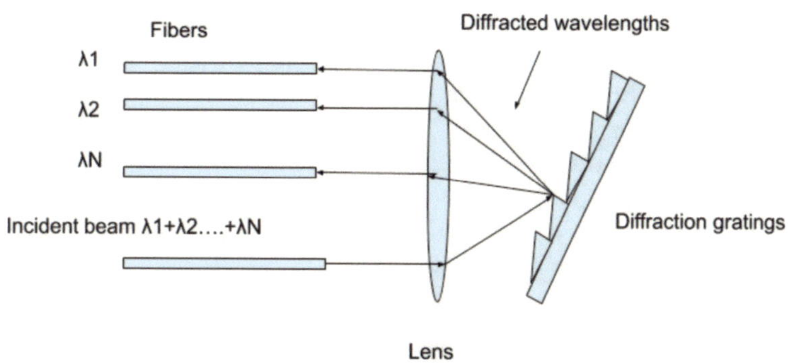

Figure 6.
Diffraction grating-based Mux configuration [10].

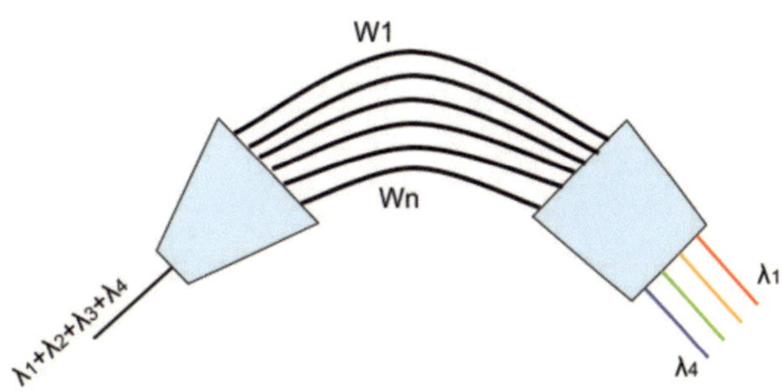

Figure 7.
AWG-based Mux/DeMux configuration [10].

- It has low losses and cross talk (insertion loss <−3 dB and cross talk <−35 dB).
- It can be fabricated on Si as photonic-integrated circuit (PIC) and can be easily integrated with photodetectors as well.

AWG suffers from drawbacks like polarization dependency and temperature sensitivity. A lot of works have been done in addressing these issues.

2.3.5 Mach-Zehnder interferometer-based devices

MZI-based Mux/DeMux works in the same principle of interference, where interfering coherent light of different wavelengths forms maxima at different spatial points and hence can be demuxed out. MZI-based Mux/DeMux devices can be integrated on silica.

2.3.6 Spectral filter-based devices

A spectral filter inserted in the optical path can be used to sort out wavelengths and hence can be used as DeMux. These devices can be implemented in different configurations, a couple of which are shown in **Figure 8**.

The first one is a fiber sandwiched at the cleaved surface of a fiber. The incident ray with two wavelengths is incident on the filter, which passes one and reflects the other. The reflected one is collected through another fiber achieving DeMux operation. Another form of filter can be implemented in a graded-index (GRIN) rod.

2.3.7 Acousto-optic filter-based devices

An interesting method for realizing a DeMux is shown in **Figure 9**. The device consists of an all-pass polarizer which linearly polarized the input signal. For demultiplexing these linearly polarized wavelengths, a combination of AOF and polarizing beam splitter (PBS) is used. The AOF can be controlled with electrical signal to rotate the polarization of a desired wavelength from transverse electric (TE) to transverse magnetic (TM). The PBS then reflects one of the wavelengths based on its polarization, resulting in DeMux operation.

2.4 Amplifier

Optical signals produced by laser, modulated with information at the multiplexer and segregated and propagated through optical fiber, are prone to attenuation and losses arising from all these components. Optical fiber technology is so advanced now that the transmission loss is practically negligible for short-haul communications. It is the component insertion loss that causes more serious signal attenuation. Eventually signal amplitude may get small enough to fall below the receiver sensitivity, which can be prevented with the use of

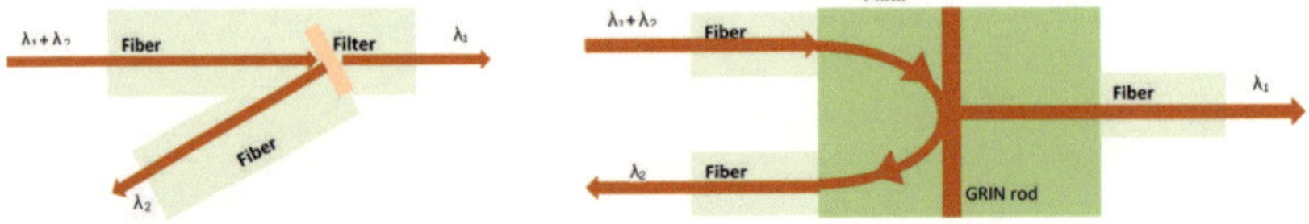

Figure 8.
Spectral filter-based devices [10].

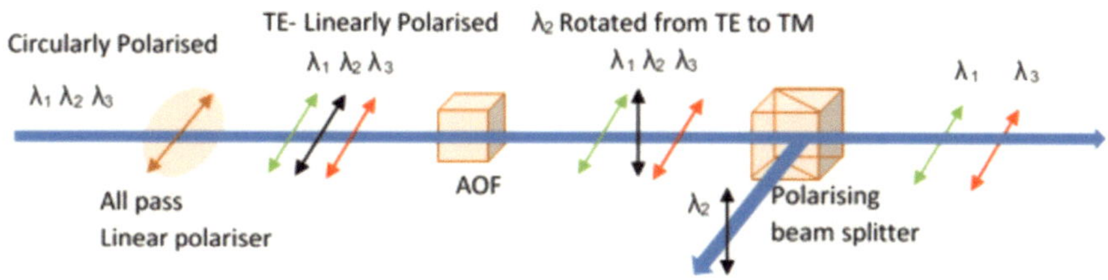

Figure 9.
Acousto-optic filter configuration [10].

amplifiers in the optical link. Before the invention of optical amplifiers, amplification could only be done in electrical domain. But this needed conversion from optical to electrical and then back to optical conversion (O-E-O). Additionally, electrical regenerators are generally sensitive to bit rate and modulation formats, which make it less flexible for additional capacity. But optical amplifiers work in optical domain and amplify the signals without O-E-O conversion. Further, they are transparent to bit rate and modulation format changes. Now, there are amplifiers that have a wide gain-bandwidth like EDFA, Raman amplifiers that can amplify signals over a large wavelength range. These facilitate the widespread use of WDM systems, which need simultaneous amplification of a wide range of wavelengths.

2.4.1 Erbium-doped fiber amplifier

EDFAs are the most commonly used optical amplifiers owing to their larger spectrum and high gain and simplicity. EDFA came into being by the early 1990s and has completely changed the landscape of optical communication industry. The most significant advantages of EDFA is its ability to amplify a wider bandwidth of signals, which is a big boost to WDM technique, as multiple channels can be amplified simultaneously. Other important aspects that make EDFA so mainstream is the availability of compact and high-power pump laser source, polarization insensitivity, easiness in coupling, absence of cross talk, and its inherent simplicity [3].

EDFA is an optical fiber with its core doped with rare earth mineral, which acts as the amplifying medium. Doping can also be done using holmium, neodymium, samarium, thulium, and ytterbium to provide gain in ranges from 500 to 3500 nm. EDFA has the capability to amplify signals in 1550-nm band, the standard telecommunication regime [11].

One main problem with EDFA is its nonuniform spectra. Different channels are amplified differently, and the difference builds up over a long-haul system with multiple EDFAs. Energy levels and gain spectrum of EDFA are shown in **Figure 10**. Several solutions have emerged in addressing this issue efficiently.

2.5 Detector

Optical detectors are devices that convert the optical signals into electrical signals. Usually a photodetector is followed by a front-end amplifier to amplify the electrical signal, which is followed by a decision circuit that estimates the data content of the electrical signal. Decision circuit needs to know the modulation scheme used for transmission. An optional optical preamplifier section can be used in front of the photodetector.

Photodetector works based on photoelectric effect. It is desirable for a photodetector to have "high sensitivity, fast response, low noise, low cost, and high

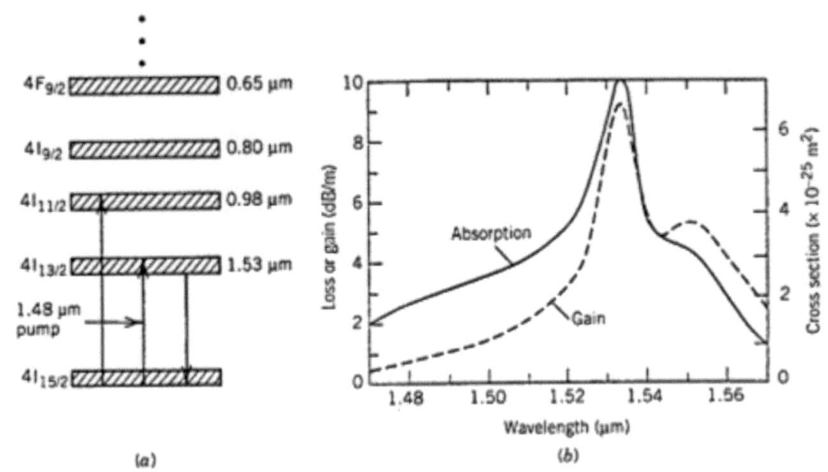

Figure 10.
(a) EDFA energy level diagram (b) absorption and gain spectra (codoped with Germania) [11].

reliability" for being more effective in communication engineering. When a photon of energy, hν, which exceeds the photodetector band gap, is incident on a photodetector, the photon is absorbed, and an electron-hole pair is generated (**Figure 11**). The electric field across the junction sweeps off this excess charge, hence producing a current flow in the external circuit.

Normally a reverse bias is applied to the junction. A reverse bias adds to the junction electric field, and the photocurrent generated by the absorption of photon is proportional to the incident optical power. It should be noted that the optical power is exponentially attenuated while it passes through the semiconductor material. The energy of the incident photon should be larger than the bandgap, e.g., of the detector. The lowest such wavelength is the cutoff wavelength above which the detector cannot operate. As Si and Ge have cutoff wavelength lower than 1550 nm, they are not used for communication application. Generally, indium gallium arsenide (InGaAs) and indium gallium arsenide phosphide (InGaAsP) are used in 1550 and 1310 nm wavelength ranges.

An important characteristic of photodetector is its responsivity, R. It is defined as

$$R = I_p / P_{in} \tag{2}$$

where I_p is the average photocurrent and P_{in} is the incident optical power. As $P_{in}/h\,\nu_e$ electrons are generated by a photon of Energy P_{in} and assuming only η fraction of incident photons is actually absorbed, then R can be written as [9]

$$R = \frac{\eta\,\lambda}{h\,\nu_e} \tag{3}$$

which can be written in terms of λ as $R = \frac{\eta\lambda}{1.24}$ where λ is expressed in μm.

2.5.1 PIN photodiode

As light is incident on the end of PN junction, the electron-hole pairs generated have to diffuse to the depletion region before getting swept away (drift) to the corresponding electrode, hence creating current in the external circuit. Diffusion velocity is slow and is in the range of 1 ns per μm. This causes the input signal to be distorted at the electrical output. Increasing the depletion region length can decrease the diffusion time. This is the idea behind PIN photodiode [3].

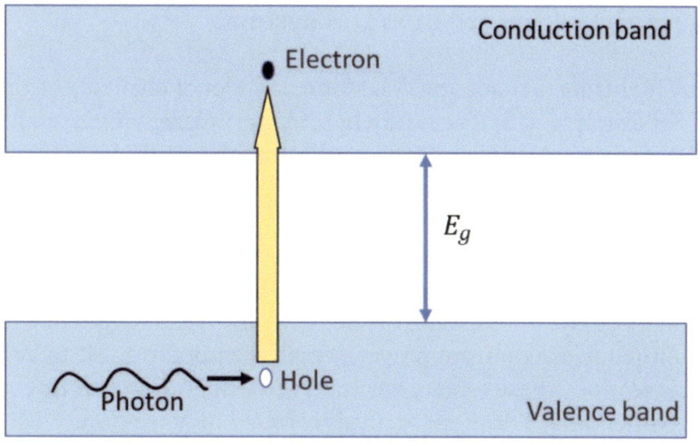

Figure 11.
Photodiode basic principle.

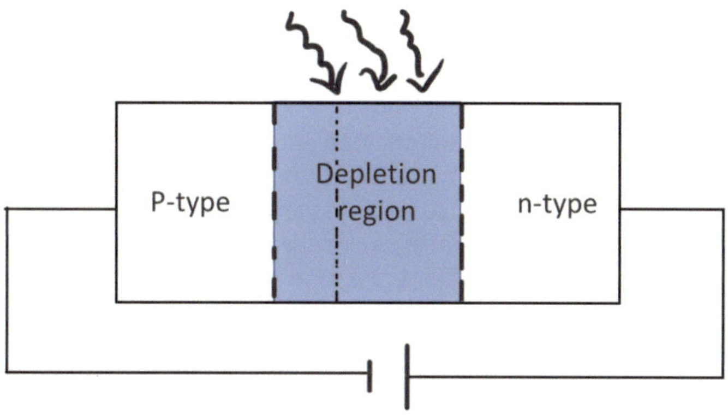

Figure 12.
PIN photodiode structure.

In PIN photodiode, the depletion region width is increased by the introduction of layer of very lightly doped intrinsic semiconductor material between p and n sides, hence the name PIN. **Figure 12** shows the structure of PN junction.

Due to the light doping, this layer provides high resistance and most of the voltage drops across it. In effect, most of the recombination happen in the depletion region; hence, the drift current far overweighs the diffusion current. So a longer W will increase the photodiode sensitivity, but a longer depletion width also implies larger transit time for the charge carries, hence increasing the response time. So there needs to be a trade-off between sensitivity and response time.

Similar to PN photodetectors, double-heterostructure design can further improve the performance of PIN photodiodes. By choosing material of sufficiently larger bandgap as p and n regions, the absorption can be limited only to the i regions. One such example is to use InP as the p and n region while using InGaP as the intrinsic layer [12]. Such a design helps to avoid the diffusion part of photocurrent, hence increasing the efficiency to nearly 100%. Reflections from the front facets can be reduced by coating with suitable dielectric layers.

Responsivity of PN diode is limited by Eq. (3). This is due to the fact that one incident photon can generate maximum of only one e-h pair. Avalanche photodiodes have internal mechanism which overcomes this and can provide larger photocurrent. They are especially preferred when the incident intensity on the photodetector is expected to be low.

3. Challenges to be addressed in WDM systems

As with any maturing technology, WDM too had a lot of challenges to be tackled as it progresses. For example, [13] discusses a hybrid multiplexer which can be used for WDM and MDM. Also WDM have found applications beyond communication, which also brings about additional challenges. Like in [14], WDM is used for optical beam steering, which is archived using photonic crystal waveguide and an integrated version of WDM in coupled micro-ring Muxs. Another field where WDM has found application is inter-chip links, as in [15] where micro-ring wavelength demultiplexers are used.

This section discusses about some of the realization challenges like EDFA transients and unequal link output power, which are not desirable in WDM-based systems. In this section, these effects are analyzed under different test cases and validated experimentally. Transient analysis is based on variations with input power, pump power, duty cycle, cascading stages, and multiplexed configurations.

3.1 Unequal channel power and its equalization

As seen before, EDFA is a major component in a WDM link for providing amplification in wavelength range around 1550-nm optical communication band. The non-flat gain spectrum of EDFA leads to uneven amplification levels for various WDM channels. This becomes more serious as more and more amplifiers are added in the link. Another serious problem for WDM networks is the wavelength-dependent gain saturation of EDFAs. Because of this effect, the loss or removal of one or more channels at the input of an EDFA can cause large changes in the output powers of the remaining channels. This effect is more predominant in CWDM systems than DWDM systems with a smaller number of wavelengths because of wider wavelength spacings when a single "C"-band EDFA is used for amplification.

To overcome this problem, an L-band EDFA in combination with a C-band EDFA can be used to flatten the EDFA gain spectrum. By using EDFA with longer fiber length or heavy doping concentration, EDFA gain characteristics can be altered [9]. So an L-band EDFA with a longer fiber length was considered. In the proposed configuration, multiplexed signal is split and passed through C- and L-band EDFAs. **Figure 13** shows the variation of gain spectra for L-band EDFA. It is suggested that if wavelengths are beyond 1565 nm, L-band EDFAs are better for practical applications [9]. CWDM configuration where the signals are splitted first and then amplified using two separate EDFAs is shown in **Figure 14**. These signals are recombined later.

3.2 EDFA transients

Due to slow gain recovery of EDFA gain, the low bit rate signals passing through EDFA undergo saturation and recovery effects during level transitions. The characteristic saturation and recovery times are, for typical operating conditions, in the range of 100 μs to 1 ms. As a result, EDFAs are intrinsically immune to the effects of cross talks at high data rates [16, 17]. The recovery time is of few hundred microseconds, hence do not affect the high bit rate signal amplification, as the erbium concentration is not significantly altered by the high bit rate signal during its short ON time period. But this is not the case with low bit rate signals, which is ON for enough time for reducing the population inversion, hence reducing the gain. The effect can be seen in **Figure 15**, when EDFA is input (1550 nm) with square optical pulse of 2 KHz and varying duty cycle, pumped with a 980-nm laser at 70 mW. The transient effects are sensitive to input signal duty cycle, signal power, and pump power of EDFA configuration. The following section focuses on EDFA transients applicable to multiplexed fiber links.

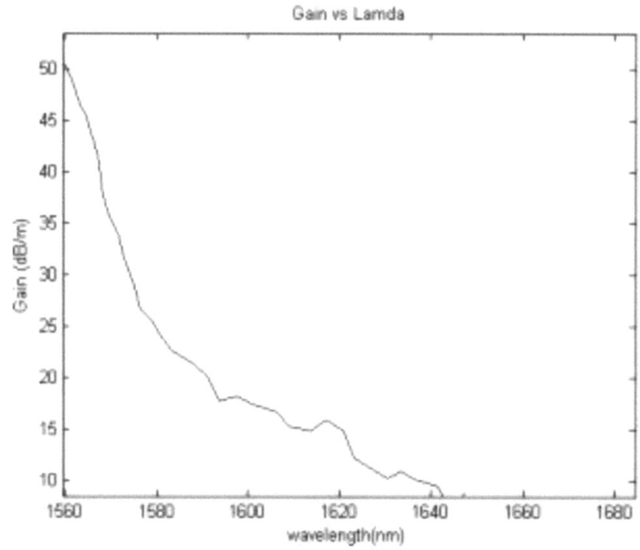

Figure 13.
L-band EDFA gain spectrum (30 m).

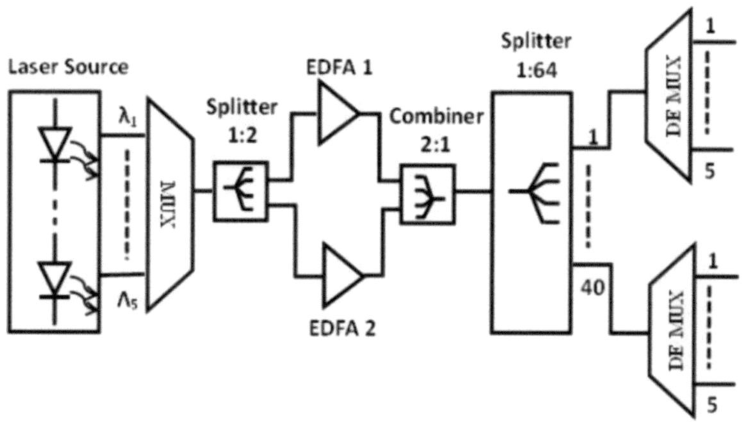

Figure 14.
Block schematic diagram with C- and L-band EDFAs for power equalization.

EDFA transients disappear with increasing bit rate as shown in **Figure 16** (bit rate 10 KHz and 1 MHz). It also decreases with lower signal and pump power (**Figure 17**).

Transient effects can produce a negative impact as the pulse shape at the output of the link is heavily distorted, it can lead to misinterpretation of data at the receiver side. Also, in cascaded EDFA applications, the transients can accumulate over length and can cause problems at detector stage. A few compensation techniques are described below.

3.2.1 Compensation using complementary pulse

To accomplish this, another complementary signal is multiplexed into the link at a different wavelength which ensures the EDFA input power remains constant. The block schematic and results are as shown in **Figure 18**. As the wavelength of compensation signal approaches the original wavelength of the signal pulse, the distortion is reduced, i.e., the closer the compensation wavelength, the better the compensation.

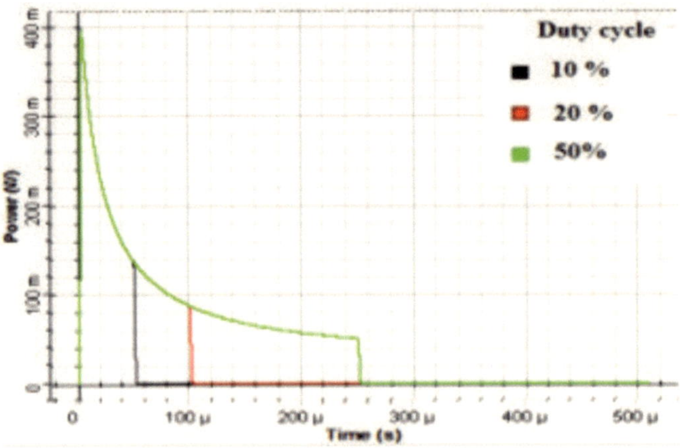

Figure 15.
EDFA response to duty cycle variation (single input pulse).

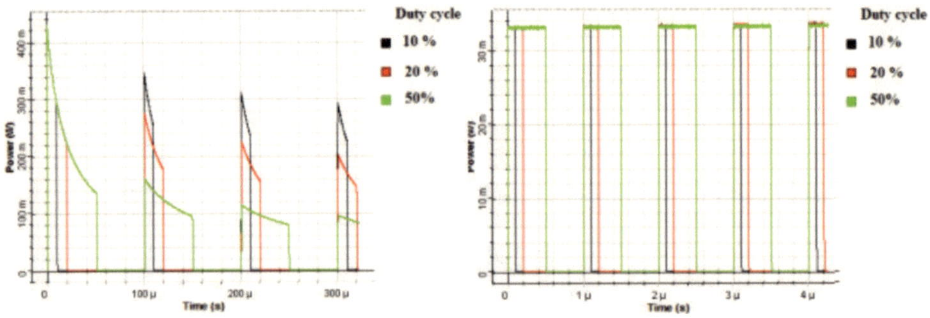

Figure 16.
EDFA response to 10 KHz (left) and 1 MHz (right) with different duty cycles.

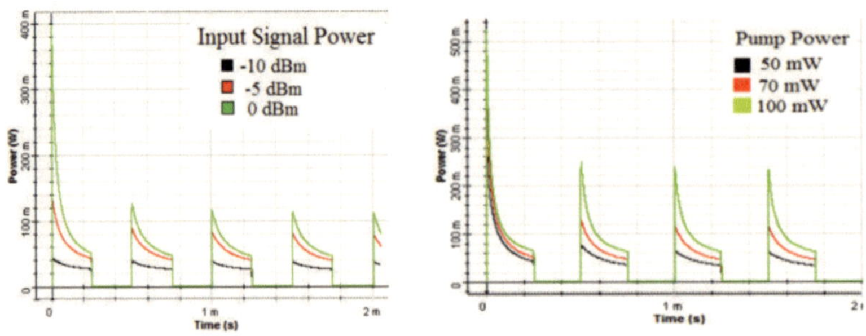

Figure 17.
EDFA transient response to signal (left) and pump variations (right).

3.2.2 Compensation using delayed pulse

In the case of 50% duty cycle pulse, two more additional suppression techniques are proposed. One uses electrical delay and the other uses optical delay. **Figure 19** shows the block schematic used for transient suppression of 50% duty cycle signal using optical delay line. **Figure 19** shows the experimental result of the transient suppressed output. The delay introduced should be equivalent to the ON/OFF time of transmitted signal. In this case, only one laser source is required. Optical delay can be introduced by using fiber spools of longer length. The delay can be applied electrically too; schematic and results are shown in **Figure 20**.

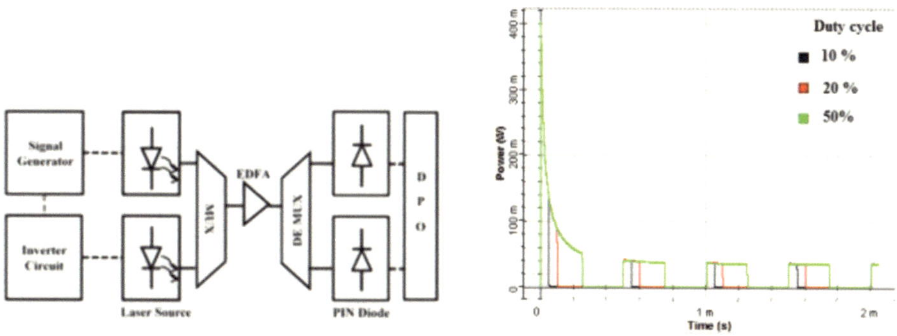

Figure 18.
Compensation using complementary pulse. Block schematic (left) and results (right).

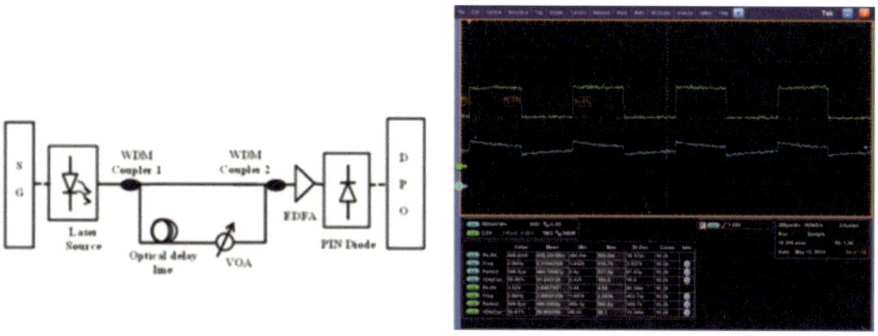

Figure 19.
Compensation using delayed pulse. Schematic (left) and experimental result (right).

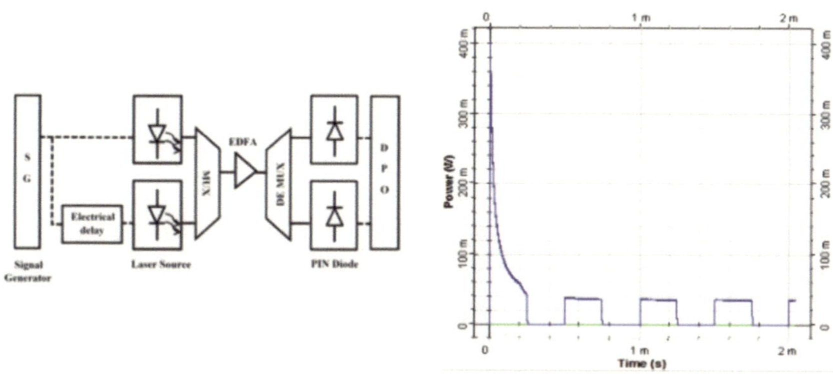

Figure 20.
Compensation using electrical delayed pulse. Schematic (left) and result (right).

EDFA transients affect WDM systems too in a similar manner by distorting the transmitted signal [17]. The description of the same is not included within the scope of this chapter.

4. Conclusion

The chapter introduces the concept of optical multiplexing with special focus on wavelength division multiplexing. Other multiplexing methods are also briefly described highlighting the operation and potential applications. A WDM link is explained by going into detail of the different components making up the link. The chapter also includes a few challenges which degrade the performance of the link and potential methods to overcome those effects.

With the WDM Mux/DeMux described above, adding or dropping an unplanned channel may require the traffic in the entire link be suspended. But with a reconfigurable optical add-drop multiplexer (ROADM), an operator can remotely reconfigure the multiplexing so that data in the other channels are not interrupted. Several technologies are developed for achieving this.

Another interesting development is the emerging of super-channels, which reduces the channel gap close to the Nyquist bandwidth. The idea is to combine multiple coherent carriers to create a unified channel, called a super-channel, which will operate at the maximum data rate supported by the analog-to-digital convertor (ADC) at the receiver. The absence of guard channels and coherent detection ensures high spectral efficiency. Some techniques include orthogonal frequency division multiplexing (OFDM), orthogonal band multiplexed (OBM), no-guard-interval (NGI)-OFDM, multichannel equalization (MCE)-WDM, Nyquist WDM, etc.

These WDM links are widely used in various regimes of communication. At present, majority of the links are made with discrete components. When a greater number of channels are required to be transmitted, a small form factor solution is preferable. Currently many researches are being carried out to bring these components to a photonic-integrated circuit form which can reduce the size to a greater extent. It is quite sure that with the latest advancements in nanotechnology, more components can be integrated resulting in a very-small-factor WDM chips.

Author details

Meena Dasan[1*], Fredy Francis[2], Kundil T. Sarath[3], Elambilayi Dipin[3] and Talabattula Srinivas[4]

1 Electronics and Radar Development Establishment (LRDE), DRDO, Bangalore, India

2 Indian Institute of Technology Madras, Chennai, India

3 Nanatom Technologies, Bangalore, India

4 Electrical Communication Engineering, Indian Institute of Science, Bangalore, India

*Address all correspondence to: dmeenasatish@gmail.com

References

[1] Space Division Multiplexing. Available from: https://www.rp-photonics.com/space_division_multiplexing.html [Accessed: June 14, 2019]

[2] Xu K, Wang R, Dai Y, Yin F, Li J, Ji Y, et al. Microwave photonics: Radio-over-fiber links, systems, and applications. Photonics Research. 2014;**2**(4):B54-B63

[3] Agrawal GP. Fiber-Optic Communication Systems. John Wiley & Sons; 2012

[4] Akiba SH, Usami MA, Utaka KA. 1.5-μm λ/4-shifted InGaAsP/InP DFB lasers. Journal of Lightwave Technology. 1987;**5**(11):1564-1573

[5] Li GP, Makino T, Moore R, Puetz N, Leong KW, Lu H. Partly gain-coupled 1.55 mu m strained-layer multiquantum-well DFB lasers. IEEE Journal of Quantum Electronics. 1993;**29**(6):1736-1742

[6] What When How. Optical-Time-Division-Multiplexed-Communication-Networks- Part-4. Available from: http://what-when-how.com/fiber-optics/Optical-time-division-multiplexed-communication-networks-part-4 [Accessed: June 14, 2019]

[7] Phillips ID, Ellis D, Thiele J, Manning RJ, Kelly AE. 40 Gbit/s all-optical data regeneration and demultiplexing with long pattern lengths using a semiconductor nonlinear interferometer. Electronics Letters. 1998;**34**(24):2340-2342

[8] Nishihara H, Haruna M, Suhara T. Optical Integrated Circuits. New York; 1989. pp. 7-8

[9] Ramaswami R, Sivarajan K, Sasaki G. Optical Networks: A Practical Perspective. Morgan Kaufmann; 2009

[10] Optical Demultiplexers. Available from: https://ftp.utcluj.ro/pub/users/cemil/dwdm/dwdm_Intro/5_5311736.pdf [Accessed: June 14, 2019]

[11] Desurvire E, Bayart D, Desthieux B, Bigo S. Erbium-Doped Fiber Amplifiers: Device and System Developments. New York, NY: Wiley-Interscience; 2002

[12] Tucker RS, Taylor AJ, Burrus CA, Eisenstein G, Wiesenfeld JM. Coaxially mounted 67 GHz bandwidth InGaAs PIN photodiode. Electronics Letters. 1986;**22**(17):917-918

[13] Tan Y, Wu H, Wang S, Li C, Dai D. Silicon-based hybrid demultiplexer for wavelength-and mode-division multiplexing. Optics Letters. 2018;**43**(9):1962-1965

[14] Ito H, Tatebe T, Abe H, Baba T. Wavelength-division multiplexing Si photonic crystal beam steering device for high-throughput parallel sensing. Optics Express. 2018;**26**(20):26145-26155

[15] Jia H, Yang S, Zhou T, Shao S, Fu X, Zhang L, et al. WDM-compatible multimode optical switching system-on-chip. Nano. 2019;**8**(5):889-898

[16] Desurvire E. Analysis of transient gain saturation and recovery in erbium-doped fiber amplifiers. IEEE Photonics Technology Letters. 1989;**1**(8):196-199

[17] Dasan M, Thodiyil S, Francis F, Elambilayi D, Talabattula S. An investigation on transient effects in EDFA with variable duty cycle and wavelength multiplexed signals. Microwave and Optical Technology Letters. 2015;**57**(2):352-356